T0122875

Bias in Science and Communication

A field guide

Bias in Science and Communication

A field guide

Matthew Welsh

Australian School of Petroleum,
University of Adelaide, Adelaide, Australia

IOP Publishing

ISBN   978-0-7503-1311-7 (ebook)
ISBN   978-0-7503-1312-4 (print)
ISBN   978-0-7503-1313-1 (mobi)

DOI   10.1088/978-0-7503-1311-7

Version: 20180501

IOP Expanding Physics
ISSN 2053-2563 (online)
ISSN 2054-7315 (print)

British Library Cataloguing-in-Publication Data: A catalogue record for this book is available from the British Library.

Published by IOP Publishing, wholly owned by The Institute of Physics, London

IOP Publishing, Temple Circus, Temple Way, Bristol, BS1 6HG, UK

US Office: IOP Publishing, Inc., 190 North Independence Mall West, Suite 601, Philadelphia, PA 19106, USA

This book is dedicated to:

My parents, for their encouragement of my early, scholarly tendencies and forbearance during my extended university studies.

Robbie and his soon-to-be little brother, in the hopes that one day they might want to read it.

And Cassandra, for one more reason every day.

(With special thanks to the baristas at the Howling Owl for their Herculean efforts in maintaining my blood-caffeine concentration.)

Contents

Part III Implications and solutions

Preface

This book is intended as an introduction to a wide variety of biases affecting human cognition, with a specific focus on how they affect scientists and the communication of science. A significant point, however, should be made up front: scientists are people and the biases that are discussed herein are, for the most part, generic in that they affect people in general rather than being specific to any particular group of people. That is, the decision making biases of experts and specialists tend to be more similar to those of lay-people than different (although chapter 10 will discuss situations where that is not the case). The role of this book, therefore, is to lay out how these common biases affect the specific types of judgements, decisions and communications made by scientists.

The book is divided into four parts. The first (chapters 1–3), introduces the reader to a variety of decision biases, the field of decision making in general and fundamental considerations regarding the psychology underlying different types of communication.

Each chapter in the second part of the book (chapters 4–10) will focus on a specific bias or a set of related decision making tendencies, describing the general effect, how they impact decisions and some of the implications for scientists' decisions and communications.

Part 3 (chapters 11–13) brings insights about these individual biases together to demonstrate how they can combine and interact to produce a variety of well-documented effects, including publication bias and stubborn denial of what, to scientists, are regarded as accepted facts. It also covers, more broadly, the ways in which biases can be overcome or avoided.

Finally, part 4 (chapter 14) draws overall conclusions about the impact of biases on science and communication, with advice on how best to move forward given what we know about their modes of action and amelioration strategies.

In all cases, an effort has been made to ensure that the latest information is incorporated and, where there are disputes or disagreements over the causes or nature of biases, alternative views are noted for those interested in following up in greater detail.

Each chapter also includes advice or exercises to help readers to identify or reduce biases in their own thinking.

Author biography

Matthew Welsh

Matthew Welsh was born in Melbourne, the second son of a second-generation banker, and grew up in Adelaide, Australia, attending Unley High School and Pembroke School.

Enrolling at the University of Adelaide with the intent of becoming an astrophysicist, he quickly discovered that university level maths was not for him and tried a number of other areas of study— branching out into classics, philosophy and genetics— before finding, in psychology, the right mix of science and art(s).

His PhD, under the supervision of Ted Nettelbeck, was a behaviour genetic study of rodent learning, as a stepping stone towards examining the contribution of genetics to intelligence—back when that sort of thing was less workaday. Research assistant work with Michael Lee on decision making, however, led to an introduction to Steve Begg and Reidar Bratvold at the (then) newly established Australian School of Petroleum (ASP).

Since that time, Matthew has been employed at the ASP on industry-funded grants examining a wide range of issues relating to the psychology of decision making as it affects the oil industry and, specifically, the scientists and engineers on whose estimates and decisions industry performance rests.

In this role, he has written more than 40 papers considering the heuristics and biases approach to decision making, the role of bounded rationality, elicitation techniques and the role of individual differences in bias susceptibility; made presentations within and taught courses to major Australian and international oil companies including Santos, Woodside, BG and ExxonMobil; collaborated with researchers at UCSF and within AstraZeneca on medical decision-making research; conducted research with defence organisations; and pretended to be a petroleum engineer and a geophysicist at major international conferences.

Matthew occasionally teaches the psychology of decision making within the Australian School of Petroleum, the University of Adelaide's School of Psychology and Flinders University's Master of Public Administration's Risk Management topic. He is a member of the Cognitive Science Society and Association for Psychological Science and a sometime member of the Society of Petroleum Engineers and American Geophysical Union.

When he is not working, he spends his time reading popular science, watching nature documentaries, writing fantastical stories and creating imaginary worlds filled with monsters ready to do battle with his (as yet) undefeated son.

This is his first book.

Part I

Introduction

Bias in Science and Communication
A field guide
Matthew Welsh

Chapter 1

Pop quiz: a battery of decision-making questions for self-assessment and reflection

It may seem strange to start a book with a quiz—a little like having students sit an exam before they have enrolled in a course. A moment's reflection, however, should reveal the benefits of doing exactly that.

Specifically, it provides us with a baseline—allowing us to determine what the examinee already knows, which is important if we want to understand what, if anything, they are learning in the course. Otherwise, when they sit an exam, we cannot be sure to what extent their mark reflects what we have taught them as compared to their pre-existing knowledge of the subject.

When thinking about how people make decisions and the biases that can affect judgements and attempted communications, this becomes doubly important. The reason for this is that we are *all* susceptible to biases that occur as a result of how our minds process information and that a number of these effects (which will be discussed in later chapters) make our memories fallible.

A particular result of this is what we call *hindsight bias*: the tendency of people to reconstruct their memory of events to include events and knowledge that they did not possess at the time. When discussing decision-making biases, for example, it is quite easy—after a bias has been pointed out—to think that you would have seen and avoided it.

The purpose of this quiz, therefore, is to act as a mnemonic aid; such that, when a bias is described, you can refer back to your answers to this quiz and see whether you displayed the bias or not, free of the effect of hindsight bias and similar effects. That is, the quiz should be used for self-assessment and reflection when later chapters refer back to and expand upon specific questions from the quiz.

1.1 Exam conditions

The decisions that are the most difficult and thus the most interesting to consider are, typically, those made under conditions of uncertainty—that is, where we possess incomplete information on the options available to us and their values. This reflects the important decisions that scientists have to make, requiring extrapolation beyond what we already know to unknown or future events or states. The quiz questions do their best to replicate this uncertainty but do, of necessity, tend to have known right and wrong answers—so that when we come to discuss the effects that these questions are designed to highlight, you can look back and see how you fared on a specific question.

This means, of course, that the quiz needs to be taken under 'exam conditions', where you attempt to answer each question based on your current knowledge and beliefs. Your ability to search the internet for answers to questions is not of interest here and doing so will only undermine the utility of the quiz.

Similarly, overanalysing the question looking for the 'trick' is of limited use as what should be of interest is how you think typically, not when specifically primed to look for biases (despite this being an understandable approach given that you have chosen to read a book on biases). When doing the quiz, therefore, it is best to put yourself under some time pressure—to replicate real-world decision making and ensure that you are making decisions as you typically would rather than overanalysing everything. (I would suggest that the quiz should take no more than 10 min to complete.)

Finally, for the quiz to act as a mnemonic aid, you will also need to record your answers, rather than just trying to remember them. Otherwise, you will be subject to exactly the effects that this quiz is designed to help you avoid.

Your 10 min starts now. Please turn the page.

1.2 Quiz

Question 1.

Imagine that you are a contestant on a game show. You have won your way through to the prize round and are now faced with selecting which of three doors to open—knowing that each has a prize behind it but that one of these is significantly more valuable than the other two. Other than that, you have no information. So, do you select A, B or C? _____

After you make your selection the game-show host opens one of the other doors to show that it has one of the less valuable prizes behind it. 'Monty' then asks you, "Would you like to keep the door that you previously selected or would you like to switch to the other door?" Would you prefer to stay or switch? _____

Question 2.

Imagine that you have put in an application for a research grant. The funding body contacts you to explain that, this year, they have only had two applications (including yours) and, as a result offer you two alternatives:

a) Accept a $50 000 grant.
b) Wait for the funding body to decide between the two grants, meaning you will receive either nothing or $100 000.

Which option would you prefer?

Question 3.

Imagine that you have access to two robotic sorting machines designed to sex *Drosophila melanogaster*. The 15-Fly Machine takes a sample of 15 flies at a time while the 45-Fly Machine accepts a sample of 45 flies. Each then accurately counts the number of male and female flies in its sample and flags the sample as unbalanced if the proportion of male flies is 60% or above.

Over the course of a week's testing, each machine sexes 100 groups of flies. At the end of the week, which machine do you think will have flagged more samples as unbalanced? The 15-Fly Machine or the 45-Fly Machine? Or will the number be approximately the same? _____

Question 4.

Imagine that you are a doctor engaged in a testing program for BRN syndrome (BRNS). In the population at large, 1 in 1000 people (0.1%) are known to have a genetic mutation causing BRNS. Your test is 100% accurate in detecting this mutation when the patient actually has it, but also has a 5% false positive rate. You have just tested a randomly selected person and their test has come back positive for BRNS. In the absence of any other information, what is the probability that the patient has BRNS?

Question 5.
Look at the following list of twelve German state capitals. Which are the six largest?

Berlin	___	Hanover	___	Saarbrucken	___
Bremen	___	Mainz	___	Schwerin	___
Dusseldorf	___	Munich	___	Stuttgart	___
Hamburg	___	Potsdam	___	Wiesbaden	___

Question 6.
Imagine you are deciding whether to replace a current testing process with a faster, computerised one. The new test is said to produce a positive result for any sample where the old test produces a positive result. You have tested two samples with each of the old and new tests, getting one positive and one negative result in each case—as shown in the table below. Which of the four samples do you need to retest in order to test the truth of the claim that the new test will produce a positive result whenever the old test does?

	Sample 1	Sample 2	Sample 3	Sample 4
Old Test	+ve	−ve		
New Test			+ve	−ve
Retest?				

Question 7.
Now imagine that you have put in an application for a research grant and received word that you have been awarded the full $100 000 that you applied for. While you are completing the paperwork to accept the grant, however, the funding body contacts you to explain that they have made an error—specifically, they have failed to assess one other, rival grant. They offer you two alternatives:
    a) Accept a $50 000 reduction in your grant amount.
    b) Wait for the funding authority to decide between the two grants, which will result in you losing all $100 000 should they select the rival grant or losing nothing if they select your grant.

Question 8.
You are looking for a person to join your trivia team and a friend of yours has suggested two of their friends—John and Jane—as possible options. You ask your friend to describe the two and she responds that John is:

    intelligent, industrious, impulsive, critical, stubborn and envious.

She then describes Jane as:

    envious, stubborn, critical, impulsive, industrious and intelligent.

Who would you rather have on your trivia team?

Question 9.
Imagine that you have won a competition and the prize on offer is a choice between *three* holiday packages.
  A. An all-expenses paid, one-week safari in Tanzania, Africa. Leaving from your home city with all transfers included.
  B. An all-expenses paid, one-week tour to the ruins of Machu Picchu in Peru. Leaving from your home city with all transfers included.
  C. An all-expenses paid, one-week tour to the ruins of Machu Picchu in Peru. All transfers are included but it leaves from a neighbouring city to which you will have to make your own way.

Which would you prefer? _____

Question 10.
How do you think you rank against your peers (e.g. people at your place of work) on each of the following traits?

		Percentile								
		$\leqslant$10th	20th	30th	40th	50th	60th	70th	80th	$\geqslant$90th
1	Driving ability									
2	Generosity									
3	Thoughtfulness									

Question 11.
For each of the following five questions, give a low value and a high value such that you are 80% confident (i.e. would accept 1:4 odds) that the true answer will lie within your range:
  a. As of January 2017, how many officially (International Astronomical Union) named moons did Saturn have?
  (   )   to   (   )
  b. What is the melting point of vanadium in degrees Celsius?
  (   )   to   (   )
  c. How many million years ago did the Jurassic period begin?
  (   )   to   (   )
  d. What proportion of the Earth's surface is covered by the Atlantic Ocean (using the International Hydrographical Organization's definition of its extent)?
  (   )   to   (   )
  e. In what year was penicillin first used to successfully treat a bacterial infection in humans?
  (   )   to   (   )

Question 12.
Looking back at the five questions immediately above, how many of your ranges do you think will contain the true value? _____

Question 13.
In Australia in 2016 14 428 people died from the three causes shown below. Please indicate how many of these deaths resulted from each cause.
    Shark attacks _____
    Road accidents _____
    Dementia _____

Question 14.
    a. Is the tallest mountain in the UK (Ben Nevis) higher or lower than 2017 m? _____

    b. How high would you estimate Ben Nevis to be? _____

Question 15.
    a. Is the longest river in Africa (the Nile) longer or shorter than 3425 km? _____
    b. How long would you estimate the Nile to be? _____

Question 16.
    a. Is the lowest temperature ever recorded on Earth above or below $-45\ °C$? _____

    b. What would you estimate the lowest recorded terrestrial temperature to be? _____

Question 17.
    a. Is North America's population more or less than 12% of total world population? _____
    b. What would you estimate North America's population to be (% of world)? _____

Question 18.
List five scientists from your field of expertise/interest.

    _____
    _____
    _____
    _____
    _____

Question 19.
List five famous scientists (from any field or fields).

_____

_____

_____

_____

_____

Question 20.
If you are running a race and you pass the person in second place, what place are you in? ____

Question 21.
A farmer had 15 sheep and all but 8 died. How many are left? _____

Question 22.
Emily's father has three daughters. The first two are named April and May. What is the third daughter's name? _____

IOP Publishing

Bias in Science and Communication
A field guide
Matthew Welsh

Chapter 2

Anchors aweigh: an introduction to the book, decision making and decision biases

Decision making is a field of research that stretches across a number of disciplines, incorporating mathematics, economic theories and psychological approaches to understanding how people should and do make decisions. This chapter gives a (necessarily brief and somewhat out-of-order) history of decision making in an attempt to explain how the current focus on the impact of cognitive biases on the decisions that people make came about.

2.1 A brief history of decision making

2.1.1 Game on

People have, of course, been making decisions since long before any organised attempt to study the hows and whys of decision making came about. Across the centuries, a great many philosophical texts have concerned themselves with aspects of decision making—as any discussion of morals must include consideration of how people make morals decisions, at the least. The type of decision making that led to economic and psychological approaches, however, has a somewhat less upstanding past.

Modern decision-making theories have their basis in probability theory, which developed as a field in the seventeenth and eighteenth centuries in Europe as wealthy patrons discussed with mathematicians (themselves often scions of well-off families) how best to maximise their chances of winning at games of chance (for an excellent, detailed discussion of this, see Bernstein's (1996) *Against the Gods*).

That is, when playing cards and dice and even betting on contests of skill, these benefactors wanted to know how to increase their own chances of winning and, in order to do so, they needed to understand how probability worked. Many of the basic tenets of probability theory sprang from these communications.

For instance, binomial probability—the branch of mathematics dealing with events having only two outcomes—came about as a result of discussions between Pascal and Fermat regarding how best to divide a wager when a context was interrupted. Specifically, if your money was already in the pot, when some event causes the game to be stopped, could you argue that you are entitled to a greater share of the money than you put it because you were more likely to win from that position that your opponent?

A simple example of this is a tennis match. Imagine that you and a 'friend' (the reason for the inverted commas should become clear as we continue) of yours are watching a tennis match and have bet $50—putting $100 in the pot. The player you predicted would win has lost the first two sets (of a five-set match) when the match is called off due to a rain event of such magnitude that the entire tournament has to be abandoned due to flooding (rather than simply delaying the match).

You reach out to take back your $50 back from the pot but your 'friend' objects that her player was clearly on track to win and that she should, as a result, get more than her original $50. In fact, she argues, given the score when the match was abandoned, she has already won the bet and wants the whole $100. You, naturally, object and your 'friend' frowns but agrees to hear you out.

Drawing on your vast knowledge of tennis statistics, you point out that, across their careers, the two players have been evenly matched, with each winning 50% of the total number of sets they have played   thereby justifying the 50/50 odds of the original bet. Given this long run average, you sketch out what might have happened had the match continued—as shown in figure 2.1.

Looking at the figure, you can see that binomial probability indicates that, despite being two sets to love down, your player still has a one-in-eight (12.5%) shot of staging a comeback. Your 'friend' agrees and you walk away with $12.50 of the original $100 pot—and one less friend!

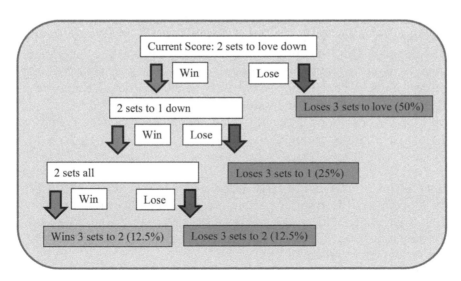

Figure 2.1. Binomial probability of tennis results with 50% chance of winning each set.

On the basis of such work, Laplace (1814) declared that probability theory was, simply, 'common sense reduced to calculus' and argued that it matched what people (or, at least clever people) instinctively did. This was the basis of the idea of *expected value*; that the correct decision to make was the one that maximised a person's expected value—as calculated from the value of the different outcomes multiplied by their likelihood of occurring.

2.1.2 Let us play a game

Before we continue our discussion of decision making, consider the following game:

A coin will be tossed repeatedly until it comes up heads. If it comes up heads on the first toss, then the player wins \$1. If the coin comes up tails, the prize doubles. That is, if the sequence is TH, then the player winds \$2, TTH yields \$4, TTTH \$8, and so on.

You are asked to make a bid in a silent auction (i.e. where each person makes their bid without knowing what other people are bidding) for the right to play this game. Given that a player is guaranteed to win at least \$1, playing is *definitely* a good decision but how *much* would you be willing to pay to play this game? Write your bid down so you do not forget it.

Now, having decided on what you are, subjectively, willing to pay to play this game, let us think about the objective value of the game. To calculate the value of an option according to probability theory, we need to calculate the value of each possible outcome. So, there is a 50% chance that the coin will come up heads on the first toss, and you will win \$1. That outcome is, therefore, worth 50 cents = 50% of \$1.

There is also, however, a 25% chance of seeing the TH sequence of tosses, which will win you \$2. Again, the value of this option is 50 cents = 25% of \$2. Then, there is a 12.5% chance of TTH, yielding \$4, which is valued at 50 cents (12.5% of \$4). You may, by now, be detecting a pattern. Each possible outcome is valued the same—50 cents. So, the total value of the game is simply 50 cents multiplied by the number of possible outcomes. This, of course is where things get interesting as the rules of the game dictate that it continues until the coin comes up heads. That is, it allows for any number of tails to be thrown prior to the concluding head—the game has *infinite* possible outcomes. Therefore, the total value of the game is \$∞.

Look back at your bid for the right to play this game. Does it mesh with the value of the game or is there a mismatch?

2.1.2.1 The St Petersburg paradox

This game is the St Petersburg paradox devised by Nicolas Bernoulli in 1713. The paradox, of course, is the mismatch between what probability theory says we should be willing to pay to play the game and the relatively small amounts that people are generally willing to pay. Certainly, it seems to undermine Pascal's claim that people are instinctive users of probability theory.

A solution to this paradox was proposed by Nicolas' cousin Daniel Bernoulli in 1738 in the form of expected utility. The insight here being that the usefulness of money does not increase in a linear fashion. Imagine, for example, the changes that winning $10 million in a lottery would bring to the average person's life. Most people would agree that this would result in significant opportunities previously unavailable to that person.

Now imagine that, a month after their first lottery win, the same person wins another $10 million. What impact on their life will this have? Again, the additional money will open up opportunities not previously available, but most people would agree that this second win will have *less* impact on the person's life than the first. That is, the second $10 million is less *subjectively* valuable. As a simple thought experiment to confirm this, consider a situation where you have just won a $10 million jackpot at a casino. The casino then offers you a chance to play double-or-nothing on a coin toss. That is, if it comes up heads, you will get $20 million but tails means you get nothing. Would you play? If not, then you are valuing the first $10 million more highly than the second as probability theory says you should be indifferent between these options as they are equally valuable.

A key insight of expected utility is that how much more valuable the first amount is, depends not just on the amounts on offer but also on the starting wealth of the person. For example, a billionaire might well be willing to play the coin toss above as the change in their overall wealth is minimal in either case. Someone who is currently broke, however, is almost certain to take the $10 million and run.

2.1.3 Homo economicus

With the addition of expected utility to probability theory, we have the basis of rational decision making as proposed within economic theories. Starting with these fundamentals, Von Neumann and Morgenstern (1944), in their book *Theory of Games and Economic Behavior*, laid out the rules for rational decision making, which have come to describe the economic view of human decision-making behaviour commonly known as *Homo economicus*—the 'rational, economic man'.

These rational rules include 'transitivity'—the idea that, if you prefer option A to option B and prefer option B to option C, then you must prefer option A to option C. Another is 'independence'—if you prefer A to B, then introducing a new option, C, will not change that preference. It also includes the idea of 'completeness': that a person has a stable preference between all available options. That is, they can state, with certainty, whether they prefer A to B, B to A, or are indifferent between the two for any two options.

This view of decision making has dominated economics since—the central idea being that people rationally make decisions so as to maximise the value of that decision to themselves. The important point here is that this was (and in some cases *is*) not simply a prescriptive theory telling people how they should behave in order to maximise outcomes. Rather, it was (is) viewed as a *descriptive* theory of how people actually behave. That is, in line with Laplace's quote (above), *Homo economicus* theory holds that people make their decisions using these rational rules and in line with probability theory and expected utility theory.

2.1.3.1 Misinformed or mistaken

Given this, the *Homo economicus* is often put forth as the normative standard against which we should measure decision-making behaviour—even by those who do not ascribe to it being a description of how people actually behave. That is, the 'rational' decision is determined and then researchers look for how and when human behaviour seems to differ from this—seeking to explain mismatches between human behaviour and predictions.

Sometimes, this turns out to be a simple case of a person being misinformed about the probability of an option or mistaken in their valuation due to limited information. This, however, is not evidence of irrational decision making so long as the person is making decisions that are rational in light of what they believe to be true. For example, accepting a bet at 7:1 odds of rolling a six on a six-sided die is a rational decision if you make the assumption that the die is fair (i.e. there is a one-in-six chance of rolling a six). If the die is loaded, then the probability of rolling a six may, however, be only one-in-ten, which makes this a losing bet. A person who does not believe the die is loaded (despite it being so), however, is not making an irrational decision by accepting the bet.

2.2 Bounded rationality

Over decades of research on decision making, however, instances of seemingly irrational decision making have been highlighted that are not so easily explained away. This field is generally traced back to Herbert Simon's (1955) paper 'A behavioural model of rational choice', which outlined a variety of problems with the *Homo economicus* model of human decision-making behaviour—specifically, the amount of information required to be 'rational' in a meaningful way.

This work led Simon to posit what he called 'bounded rationality'—the idea that our ability to act in the fully rational manner prescribed by *H. economicus* is bounded by both human cognitive limitations *and* limitations in the environmental information structure (Simon 1956). The first idea, cognitive limitations, is quickly grasped. For example, human working memory is generally held to be limited in terms of the number of objects, items or relationships that can be simultaneously recalled to around seven items (Miller 1956). As a result, anyone trying to decide amongst even a small number of options cannot simultaneously consider all of the preference relationships between these.

The second of Simon's 'bounds' is sometimes harder to grasp, but depends on the fact that the amount of information that can be gained from investigating the world is limited by the structure of the world itself. Imagine, for instance, that you were interested in undertaking a study of the variety of species of fish found within a particular lake. If you take ten samples from different locations around the lake and each of these contains the same ten species, then you could be fairly confident in having adequately characterised the lake's inhabitants. If, in contrast, each sample had turned up one distinct species, then you would be justified in being far less confident about how many other species you might find and, thus, need to continue sampling.

Simon, and researchers like Gerd Gigerenzer, who have continued this tradition, argue that these bounds make meeting the requirements for rational decision making too hard for it to be a reasonable description of what people do or even should do (see, e.g. Gigerenzer and Brighton 2009). That is, given that we are unlikely to be able to get enough information from the world to adequately characterise, for example, the value and probability of the different outcomes—or even, in many cases, to even enumerate all of the different outcomes—and that we do not have the necessary cognitive processing to combine even the limited information that we do have, we cannot reasonably expect people to act as *H. economicus* predicts and directs.

Instead, adherents of this approach argue that our decisions need to be understood in light of these limitations and that, in spite of these, we actually make very good decisions because our cognitive abilities have developed to take advantage of the world's environmental information structure. Simon described this using the two blades of the scissors as an analogy—either by itself produces poor outcomes but, combined, they produce good results because they fit together so well. It is, they argue, just when we have a mismatch between the blades that we run into problems.

2.2.1 Heuristics and biases

A key insight from the bounded rationality approach is that people do not make decisions using the rational, logical processes described by the *H. economicus* model. However, a significant body of research still relies on the normative predictions of rational decision making—comparing and contrasting these with observations of human behaviour. The discrepancies between these two are commonly described as *biases* and researchers seek to explain why these biases occur, and when and how they do as a result of the cognitive processes that give rise to them. This work suggests that, in many cases, our decision processes are far simpler that rational theories predict, being heavily reliant on simple rules-of-thumb—or *heuristics*.

Tversky and Kahneman's (1974) seminal paper 'Judgment under uncertainty: heuristics and biases' raised awareness of this approach, which has since grown to dominate the field of judgment and decision making. In this paper, the authors highlighted a variety of instances when people's decisions systematically differed from predictions made by *H. economicus* and outlined the heuristic processes they believed accounted for this. The broad direction of this research has been to identify the situations that result in people's decisions being biased, in the hope that understanding this will enable us to avoid or mitigate those biases.

Over decades of research, dozens of these heuristics and accompanying biases have been identified—including the eponymous 'anchoring' effect.

2.2.2 Two systems of thinking—an aside

Anyone taking even a passing interest in decision making will soon come across mentions of 'two systems' or 'dual process' approaches or theories. These theories—eloquently explained in Kahneman's (2011) *Thinking, Fast and Slow*—argue that people engage in two distinct forms of reasoning: system 1 is fast, intuitive thinking that occurs primarily at a subconscious level and system 2 is slow, logical reasoning that occurs at the conscious level.

This has, to some extent, come about as a result of the work done in the heuristics and biases field—mirroring the division therein between the simple, often unconscious, heuristic decision processes that people often use and the conscious, rational decisions that people could (and sometimes do) make via the application of their reasoning. This concept has a great deal of face validity—that is, it corresponds with how it seems to an observer that people think—and is probably the majority view amongst researchers interested in decision making. As such, the two systems approach will be taken as standard throughout this book, unless noted otherwise.

That said, it is certainly not the only view. Some authors, for various reasons, argue that a single reasoning system actually makes more sense and accords better with some experimental results (see, e.g. Osman 2004). Additionally, even amongst its adherents, it is widely acknowledged that 'two systems' is, at best, an oversimplification. For instance, Keith Stanovich, an originator of the system 1 versus system 2 nomenclature—in Stanovich and West (2000)—now refers to system 1 as TASS (The Autonomous Set of Systems) to reflect the fact that system 1 is not a single underlying process but rather includes any of a large number of unconscious processes (Stanovich 2009).

In the same paper, Stanovich puts forwards an argument for there being perhaps three systems rather than two but, even here, it is worth keeping in mind that the two (or three) systems approach does not incorporate the necessary metacognitive processes required for error recognition and switching between the two systems. That is, while distinguishing between the different types of reasoning that people engage in and the implications of this for the estimates and decisions that people make, the theories are largely silent on *when* and *why* people use one rather than the other. As such, while useful as a scaffold around which to organise discussion, it should always be understood to be short-hand for the far more complicated truth.

2.3 Conclusion

The abbreviated history of decision making given above makes clear the current focus of research into judgement and decision making. The inadequacies of rational, economic models to explain the majority of heuristically driven human behaviour, combined with the observation that people are, in some cases at least, capable of making better decisions through the application of reasoning, led naturally to a comparison between these approaches. This comparison of intuitive and rational decisions led to both the heuristics and biases approach to decision making and the accompanying two systems models.

Taking this view as a starting point, each of the following chapters will focus on a small number of related effects—caused by idiosyncrasies in human decision making —and explaining how and why they affect the decisions that we make and discussing whether and how their influence can be limited in cases when we need to make better decisions or more clearly communicate our opinions and findings to others.

References

Bernstein P L 1996 *Against the Gods: The Remarkable Story of Risk* (New York: Wiley)

Gigerenzer G and Brighton H 2009 Homo heuristicus: why biased minds make better inferences *Top. Cogn. Sci.* **1** 107–43

Kahneman D 2011 *Thinking, Fast and Slow* 1st edn (New York: Farrar, Straus and Giroux)

Laplace P-S 1814 *Essai Philosophique sur les Probabilites* (F W Truscott and F L Emory, Trans.) (New York: Dover)

Miller G A 1956 The magical number seven, plus or minus two: some limits on our capacity for processing information *Psychol. Rev.* **63** 81

Osman M 2004 An evaluation of dual-process theories of reasoning *Psychonom. Bull. Rev.* **11** 988–1010

Simon H A 1955 A behavioral model of rational choice *Quart. J. Econ.* **69** 99–118

Simon H A 1956 Rational choice and the structure of the environment *Psychol. Rev.* **63** 129

Stanovich K E 2009 Distinguishing the reflective, algorithmic, and autonomous minds: is it time for a tri-process theory *In Two Minds: Dual Processes and Beyond* ed J Evans and K Frankish (Oxford: Oxford University Press) pp 55–88

Stanovich K E and West R F 2000 Individual differences in reasoning: implications for the rationality debate? *Behav. Brain Sci.* **23** 645–65

Tversky A and Kahneman D 1974 Judgment under uncertainty: heuristics and biases *Science* **185** 1124–31

Von Neumann J and Morgenstern O 1945 *Theory of Games and Economic Behavior* (Princeton, NJ: Princeton University Press)

IOP Publishing

Bias in Science and Communication
A field guide
Matthew Welsh

Chapter 3

On message: reasons for and types of communication

This book is written for scientists and other technical professionals with multiple goals in mind. In addition to the goal of educating readers on the nature of biases, it is also designed to act as a field guide—giving the necessary knowledge to identify biases in the wild, so to speak. That is, identifying when information that is being communicated is likely to be affected by biases—from either the communicator or the listener.

In order to determine this, however, we need to understand *why* people are communicating and the different ways in which they do so, as these will affect the biases that are likely to be seen. This is because the motivations of communicators add another layer of potential biases on top of those resulting from our cognitive limitations. Specifically, because communication always has a purpose, this means that people use communication to achieve specific ends. If that end is not the clear transmission of information, then it becomes likely that when or how people choose to communicate (or omit) information can be a deliberate attempt to influence or bias the receiver.

Herein, we distinguish between these types of bias as 'cognitive biases' when they result primarily from people's natural, cognitive tendencies and 'motivational biases' when the person is aware and deliberately trying to introduce bias in order to gain some advantage. (Of course, the distinction is not a clean dichotomy and differences between these types of biases can be subtle and a variety of effects that we will discuss later seem to straddle the boundary.)

In the following sections, we discuss the different forms of communication that scientists need to engage in and how these are affected by biases.

3.1 Scientific communication

These are the communications with which readers will likely be most familiar. Scientific papers and books serve to present theories about how the world works, arguing from how well the theory accounts for observed facts—as do conference presentations and even personal scientific communications. As such, barring deliberate malfeasance, we might hope that such works would be free of bias. The truth, however, is that—like all forms of communication—scientific writing serves multiple purposes, which can result in the introduction of various biases.

In addition to clear cases of conflict of interest—like the MMR vaccine and autism controversy where Andrew Wakefield is alleged to have been paid hundreds of thousands of pounds by lawyers who were seeking evidence that the vaccine was dangerous in order to aid a lawsuit (Deer 2006)—there are motivational biases that are subtler in nature. Scientists do not work in isolation, which means that they are often placed in competition with other scientists for limited resources (jobs, grants, etc). This could result in pressure on an industrial scientist, for instance, to interpret findings in the most positive light given their employer's goals—particularly if promotion is contingent on conducting successful studies.

Similarly, for scientists working at universities, there is pressure to publish early in order to meet performance indicators and beat rival scientists to the discovery rather than waiting for additional, confirmatory results. Again, promotion is often tied to the rate of publication.

Finally, given the scientist's role as an 'expert', there are reputational motivators to consider as well. When asked for an opinion, the addition of caveats can undermine the authority of a statement and, as such, a scientist may be tempted to give a more definitive answer than their data actually support in order to portray their expertise.

These and other biases likely to affect scientific communications are discussed in depth in later chapters. For example, we know that: people tend to weigh evidence according to how neatly a causal explanation fits rather than how much data supports it (chapter 4); they have a natural tendency to seek confirmation of what they already believe (chapter 5); and there are a variety of effects that impact on rates of scientific publication, with the result that the literature itself can be biased (chapter 11).

A key thing to remember, however, is that, while the central goal of scientific communication is for scientists to communicate facts and theories to other scientists and technical experts, these communications are also shared with non-scientists and, as such, we need to be aware of how other people will interpret the content and message.

3.2 Personal communication

Personal communication, for example, is governed by norms and expectations regarding *how* people will communicate and thus has implications for how people interpret communications. Grice's (1989) 'maxims of conversation', for instance, hold that communication should be: as informative as required and no more (information); truthful and evidenced (quality); relevant (relevance); and both unambiguous and clear (manner).

Looking at these maxims, it is easy to see that they correspond with what we would want from good scientific communication—as well as good conversational practice. The problem, however, is that different standards are applicable to different areas and these standards are generally much higher for scientific communications than in normal conversations. The result of this is that people receiving communications from scientists can interpret the additional informativeness and clarity of expression required in precise scientific communication as undermining the strength of any statement. That is, the inclusion of all of the caveats and addenda required to make a precise, scientific point that is truthful and evidenced from the scientist's point of view would not, typically, be required in conversation. As a result, when they are included, the listener interprets this as meaning that the scientist is prevaricating and does not strongly hold the opinion that they are putting forward.

Another key difference between scientific and personal communication that leads to biases in the transfer of information lies in the types of evidence that are considered convincing. In scientific communications, while the clarity of argument is still important, the strength of evidence is often highlighted statistically. Particle physicists, for example, hold off declaring a discovery until the experimental data reach the 5-sigma criterion (which can be thought about as a 1-in-3.5 million chance of seeing a trend as strong as is actually observed in the data if there were, in fact, no real effect). The same, null hypothesis significance testing approach is used in psychology, with p-values indicating the probability of seeing as strong an effect in the data just by random chance—that is, if there is no relationship between the experimental manipulations and outcomes. Convention sets the minimum p-value that is considered sufficient to reject the null hypothesis at .05, but statistically literate people know that the lower the p-value in the data, the more reliable (or replicable) the results are likely to be.

In contrast, the evidence used in personal communication is far less likely to be statistically based and statistical inference is, if anything, considered an unreliable source of evidence. Consider, for example, the quote attributed (probably falsely) to Benjamin Disraeli by Mark Twain and those following:

'There are three kinds of lies: lies, damned lies and statistics.'
    '98% of all statistics are made up.'
'Statistics can be made to prove anything—even the truth.'

While said in jest, these sayings reflect an essential truth about the nature of evidence. A typical person puts far less weight in statistical evidence than does a scientist. Part of this, doubtless, results from a misunderstanding of how statistics should actually be interpreted and, thus, what they mean (see, e.g. chapter 4) but it is also the case that statistics (or, perhaps, things that sound like statistics) are regularly misused such that a person, on hearing a statistic being quoted, is right to be suspicious.

The essence of the problem here is one of trust. Who should the listener trust and why? For the most part, statistics are impersonal and, thus, unless a person has good reason to trust the person or institution that has produced them, they will be taken with a grain of salt. Personal communications from a trusted associate, in contrast, come with a built-in level of trust—regardless of the evidence therein.

For example, imagine a person diagnosed with cancer. Their GP refers them to a specialist who proposes a course of treatment based on the latest medical interventions with attendant dangers and side-effects. The patient asks about alternative treatments and the specialist explains that the latest evidence is that people who select to be treated with alternative medicine rather than the conventional medical treatment are, on average, twice as likely to die within five years (Johnson *et al* 2018). The patient is then discussing this with a friend, who says that his mother had cancer and was cured by changing her lifestyle—without any side-effects.

A scientist, looking at these exchanges, might struggle to understand why the person then selects to forgo conventional medicine. There are, however, two clear reasons for doing so. The first is the nature of trust, with people being demonstrably sensitive in terms of how much weight they will put on data from more or less trusted sources (Welsh and Navarro 2012). Here, the patient has two sources of information. The first is received, second-hand, via someone they have just met and from a source with which they are completely unfamiliar. The second is received from someone they know and trust to have their best interests at heart and who has, presumably, directly seen their mother's recovery.

Additionally, on either side of the debate, they have *one* opinion—as the statistical evidence is, conversationally at least, subsumed into the specialist's opinion rather than counting as hundreds of separate data points.

When this is combined with the biases discussed in chapter 7, which describes the tendency of people to regard themselves and their circumstances as special cases, this results in the patient being far more likely to disregard the medical advice than is warranted by the available evidence.

The message for communication here (and expanded on in chapter 10) is clear. If you want your information to be accorded appropriate weight, you need to meet the expectations of the listener—rather than relying on the presentation of information that would convince another scientist. In the example above, for instance, the specialist would be far more likely to convince the patient to undergo treatment with an anecdote—about a single, specific patient who forwent treatment to pursue alternative approaches before returning, too late, for medical intervention—than with any number of statistical facts.

3.3 Policy communication

The above example leads naturally into the consideration of what have become known as 'nudges'. That is, ways of designing communications so as to prompt specific behavioural responses. The processes by which this can be achieved and the reasons an organisation might wish to do so are laid out clearly in Thaler and Sunstein's (2008) book *Nudge: Improving Decisions About Health, Wealth and Happiness* and work based on the underlying theories has been used by the

Behavioural Insights Team set up by the UK Government and now operating as a consulting firm with the UK, USA and Australian Governments.

The central idea here is given as 'liberal paternalism'. That is, leaving people free to choose whichever of the available options they prefer but making the one that is most desirable, from a social point of view, the default choice. The classic example revolves around organ donation and follows from the work of Johnson and Goldstein (2003), which showed that organ donation rates were significantly (massively, in fact) higher in countries where people, when filling in their car licence information, were asked if they would like to opt out of being a donor rather than opting in at the same stage.

This approach draws heavily on the heuristics and biases approach mentioned in chapter 2 and on which many of the examples in later chapters are based. That is, those in charge of drawing up forms and designing the questions to be answered rely on known cognitive biases in order to prompt more people to respond in a 'better' manner. For instance, prospect theory (described in chapter 5) makes clear that people value losses and gains quite differently and so, describing a situation in a way that makes it sound more like a loss, for example, changes how people respond—making them more willing to take risks, for instance.

The point of this for scientists is that, in cases where scientific work is being used to inform policy, we cannot rely simply on the benefits of a proposed course of action to convince people to take up the intervention. Rather, attention needs to be paid to how the information is structured and what people's natural, decision-making tendencies will prompt them to do, as people's decision making under uncertainty is volatile, reacting strongly to what seem to be minor, surface-level changes to the information presented.

3.4 Group communications and decisions

As noted above, much of the work herein is based on the heuristics and biases approach and this has been, primarily, focused on individual decision makers. That is, researchers have tended to look for general, cognitive processes that people share and that affect the decision that an individual will make when presented with a particular type of problem.

This, however, misses an important source of potential bias in decision making—group behaviours. In the above sections, we have touched on motivational and cognitive biases that affect how individuals present and interpret information but it is important to realise that these can be compounded when people are placed in groups. For example, it has long been recognised (see, e.g. Ingham *et al*'s (1974) replications of Ringelmann's (1913) study) that individuals put in less effort when working in groups than when working alone—an effect called social loafing. Similarly, group decision-making processes tend to quash alternative viewpoints and reach overconfident conclusions; an effect described as 'groupthink' (Janis 1972).

These effects will be expanded on in chapter 9—along with a discussion of how stereotypes and group identification can bias how we view information or interpret communications—but, for now, it is important to just keep in mind that group situations often act as a magnifier of motivational bias. If, for example, a person wants to avoid looking foolish in their communications (as most of us do), then the number of people to whom they are communicating is an important consideration because it increases the likelihood that someone will disagree or know something that the speaker does not. This may prompt a less confident person to simply not speak up in a group—contributing to groupthink.

3.5 Elicitation of uncertainty

All of the above aspects of communication are relevant to the problem of the *elicitation of uncertainty*—a phrase that demands some explanation. Elicitation, as used when discussing decision making, refers to the process by which we elicit information—generally from experts and, commonly, in the form of estimates about uncertain or unknown parameters (see, e.g. Kadane and Wolfson 1998). That is, it refers to the process undertaken when we are trying to reduce our own uncertainty about some element of the world by asking someone else for their insights into it.

This is, of course, most interesting and important in those situations where the expert does not *know* an answer but rather has to make educated estimates regarding the likelihood of future events or unknown states of nature—making decisions under uncertainty.

Scientists, of course, need to act as both elicitors—asking others for assistance in determining appropriate inputs to models, for instance—and also as elicitees, providing information to reduce other people's uncertainty. Given this, it is very important to understand how people construct estimates under uncertainty and how they will generally communicate such information.

An essential point, that will become clear as this book progresses and examples of human decision making and communication are discussed, is that people's estimates and even preferences are *constructed* in response to the questions that are asked of them and, thus, the manner in which these questions are posed can have a far more significant impact on the estimate that is produced than most people typically assume. Given that the role of elicitation is to reduce uncertainty, its goal needs to be the production of an unbiased reflection of the elicitee's knowledge regarding the parameter of interest. As such, understanding the ways in which elicited estimates can be biased by different question formats and people's inherent cognitive processing tendencies is a very important aspect of science and science communication.

3.5.1 Individuals in crowds

An important observation relating to how and whom we should ask for estimates is the 'wisdom of crowds' effect (Galton 1907). This is, quite simply, the observation that the mean or median of a set of independent estimates tends to be better than a randomly chosen member of that set. That is, if you want a better estimate, you can ask multiple people and average their responses rather than trying to determine who the single, best person to ask is. A key aspect of this—which relates back to the idea of groupthink raised above—is, however, its requirement for independence of people's estimates.

The wisdom of crowds is, primarily, a mathematical effect resulting from the fact that, while each person you are asking for an estimate may be biased in some way, the way in which they are biased *differs* from how the next person will be biased and thus these biases are, largely, random and averaging tends to remove them. If, however, you are eliciting values from a group of experts after they have sat down and discussed their opinions, then they are likely to have been exposed to the same potential biases and thus their estimates will not be independent and the mean or median estimate of the group will include these, systematic, biases. That is, contrary to people's expectations, the best way for a group to operate is often as a set of individuals. Too much interaction between group members decreases the benefits of getting multiple view-points and results in the overconfident assessments typical of groupthink.

3.5.3 Crowds in individuals?

The reason that the wisdom of crowds is not used as widely as its efficacy suggests it ought is, however, more than just people's tendency to expect more interactive groups to produce superior results. Exacerbating this problem is the fact that the wisdom of crowds relies on numbers to work—that is, the more people you can ask the greater the improvement should be. In many situations facing scientists and engineers, however, the number of people who can meaningfully respond to a question is restricted—by expertise or by confidentiality concerns—with the result that the effect's potential is limited.

Strangely, however, the observation regarding the nature of elicitation made in section 3.5 means that, even when we have only one expert to elicit an estimate from, we can *still* benefit from this effect. This is because elicitation is a constructive process—that is, people are not just giving you answers that they already have in their heads. Rather, they are, when asked, constructing an estimate. This means that, if the person does not remember their previous estimate or has been asked about it in a different way, they are likely to produce a somewhat different estimate each time—as a result of the limitations of human memory and changes in context leading to them being biased in different ways each time. That is, there is some degree of independence within the set of estimates produced by a single person and these can be averaged out.

Thus, asking a person to make the same estimate twice and averaging these tends to produce a better estimate than just asking them once (see, e.g. Vul and Pashler 2008) so long as independence in the estimates can be maintained in some way—by increasing the duration between estimates, for instance, or by other means (see, e.g. the description of the MOLE elicitation technique in chapter 7).

3.6 Conclusion

Clear communication is a keystone of science and a central aim of this book is to assist scientists in seeing instances where their own communications could be misinterpreted or unconsciously biased and where information being presented by others may have been tweaked (deliberately or subconsciously) to serve the communicator's purposes rather than just to transmit knowledge.

The following chapters present specific examples of heuristics and biases—some cognitive and some motivational—all of which have the potential to affect the clarity of estimates elicited from experts and communicated information and, thus, bias the decisions that people are likely to make.

References

Deer B 2006 MMR doctor given legal aid thousands *The Sunday Times* 31 December https://web. archive.org/web/20070223155047/http:/www.timesonline.co.uk/tol/newspapers/sunday_times/britain/article1265373.ece (Accessed: 23 September 2007)

Galton F 1907 Vox populi (the wisdom of crowds) *Nature* **75** 450–1

Grice H P 1989 *Studies in the Way of Words* (Cambridge, MA: Harvard University Press)

Ingham A G, Levinger G, Graves J and Peckham V 1974 The Ringelmann effect: studies of group size and group performance *J. Exper. Social. Psychol.* **10** 371–84

Janis I L 1972 *Victims of Groupthink; a Psychological Study of Foreign-policy Decisions and Fiascoes* (Boston, MA: Houghton, Mifflin)

Johnson E J and Goldstein D 2003 Do defaults save lives? *Science* **302** 1338–9

Johnson S B, Park H S, Gross C P and Yu J B 2018 Use of alternative medicine for cancer and its impact on survival *J. Natl Cancer Inst.* **110** 121–4

Kadane J and Wolfson L J 1998 Experiences in elicitation *J. R. Statist. Soc.* D **47** 3 19

Thaler R H and Sunstein C R 2008 *Nudge: Improving Decisions about Health, Wealth, and Happiness* (New Haven, CT: Yale University Press)

Vul E and Pashler H 2008 Measuring the crowd within: probabilistic representations within individuals *Psychol. Sci.* **19** 645–7

Part II

Biases in judgement and decision making

IOP Publishing

Bias in Science and Communication
A field guide
Matthew Welsh

Chapter 4

Improbable interpretations: misunderstanding statistics and probability

The focus of this chapter is on some of the ways in which people interpret numerical information and, specifically, the ways in which this tends to differ from the dictates of probability theory. That is, we begin with the evidence that Laplace's view of human reasoning—equating probability theory with common sense—is far too optimistic. In contrast, people's grasp of probability theory and the statistical analyses that depend on it is often poor and biased in systematic ways.

In addition to this, we will touch on questions of definition and usage of common terms related to probability and decision making that affect how people understand the information presented to them.

4.1 Risk, variability and uncertainty

Thus, before going any further we should take a minute to ensure that we are on the same page regarding some fundamental terminology relating to probability. Specifically, the terms: risk, uncertainty and variability. This is important as these terms are used in different fields to mean quite different things and thus present a possible area of confusion or miscommunication.

4.1.1 Risk

For example, 'risk' is used in finance to refer to the volatility of a commodity—that is, it is a measure of how variable it has been in the past or how much uncertainty there is regarding its future values. In petroleum exploration, in contrast, 'risk' is often used to describe the probability of a negative outcome such as a well not finding oil. In risk management, it is used differently again—with a 'risk' being the product of an event's likelihood and magnitude. Additionally, we have 'risk' used as a descriptor of people's behaviour—with people being called 'risk seeking' or 'risk averse'. Finally, we have the common usage of the term 'risky' as a synonym for two separate concepts: dangerous and uncertain.

These regularly cause confusion. For example, the term 'risk seeking' is used in common parlance to refer to a person who enjoys taking part in activities that others regard as dangerous—sky-diving, parachuting, etc. In the decision-making literature, however, it has a far more specific meaning—relating to a person's decision in so-called 'risky choice' paradigms (see the description of 'framing' in chapter 6) that involve no risk to a person's body but rather a decision between a certain option and an equivalently valuable but uncertain option. For this reason, wherever possible the term 'risk' will be avoided and instead, decisions will be discussed in terms of their possible outcomes (positive or negative) and the probabilities of these occurring.

4.1.2 Variability

In contrast to risk, variability is a far clearer concept, referring to cases where measurements of the same parameter vary across time or space. For example, daily temperatures are variable. We can measure the same parameter—e.g. temperature in degrees Celsius—at a variety of locations at the same time or at the same location repeatedly and, in either case, we end up with a set of measurements that differ from one another. That is, they are variable.

Thus, variability is easily understood and has a common definition across fields. Where it can cause confusion, however, is where it is not clearly distinguished from uncertainty—as outlined below (for a more detailed discussion of this, see Begg *et al* 2014).

4.1.3 Uncertainty

Uncertainty, like risk, is defined somewhat differently in different fields—with some economists following Knight (1921) in defining uncertainty as 'things that cannot be measured' in contrast to 'risks' which can. More typically, however, (such as in decision science) 'uncertainty' is used to refer more broadly to not knowing whether a statement is true or not—or not knowing what value an as-yet-unobserved parameter might take (see, e.g. Bratvold and Begg 2010).

This is where it runs the risk of confusion with variability. If, for example, we wanted to predict the average, global temperature in 2025 then this is our uncertain parameter—as we do not know what the value will be. What we might have, however, is measurements of the variability in the past 100 years' average global temperatures. A common error is for people to assume that the past variability is an adequate measure of possible future values. That is, they use the variability as their measure of uncertainty.

That this is an error is made clear by recent temperature trends—with each of 2014, 2015 and 2016 being hotter than any year in the preceding records (NASA 2016). That is, each of these observations lies outside the range indicated by prior variability.

Our uncertainty regarding future temperature measurement, needs to take into account what we know—not just about the variability of previous data but also the trends in those data, the effects of current human activities on the atmosphere and ocean, and the probability of policy decisions or new technologies changing those activities and/or their effects.

4.2 Subjective probability

That is, uncertainty is subjective and, as a result, the probabilities we should assign to different outcomes depend on what we (individually) know about the event— rather than being objective in the way that variability should be. This is relatively easy to demonstrate but many people still struggle to draw the distinction. A key example of this is the Monty Hall problem (Selvin 1975), named for the host of the American game show 'Let's Make a Deal', and which is demonstrated below.

4.2.1 Monty Hall

Imagine that you are a contestant on a game show. You have won your way through to the prize round and are now faced with selecting which of three doors to open— knowing that each has a prize behind it but that one of these is significantly more valuable than the other two. Other than that, you have no information and so, in accordance with probability theory, you simply guess—selecting a door at random. So, do you select A, B or C?

After making your selection, however, the game-show host opens one of the other doors to show that it has one of the less valuable prizes behind it. 'Monty' then asks you, 'Would you like to keep the door that you previously selected or would you like to switch to the other door?' Which would you prefer? To stay or switch? Decide before reading on.

The majority of people, faced with this choice, elect to stay with their original choice and their justification for doing so is that the probability of their door having the better prize behind it is the same as the other door. That is, having gone from three doors down to two doors, they believe that the probability of them having the winning door has changed from 1/3 to ½ and so switching makes no difference.

This reasoning, however, is flawed as it ignores the additional information that becomes available when the host selects a door to open. Specifically, that the host *knows* where the major prize is and, when opening a door will, therefore, avoid that door. 'Monty' is also limited in that he cannot choose to open the door that you have selected. If you have selected the winning door then that is no problem for the host as he still has two doors to select from. If, however, you have selected *either* of the wrong doors, then 'Monty' has to open the only other wrong door—leaving the major prize hidden.

The implication of this is that there is a 1/3 chance that staying with your original door will win you the major prize—as only when you have originally chosen the winning door when the probability was, in fact, one-in-three, will the remaining door have a minor prize. If the prize is behind either of the doors that you did not select, which has a 2/3 probability, then switching after Monty reveals the minor prize ensures that you will get the major prize as that is the only door that can remain. Thus, switching gets you the major prize 2/3 of the time—as shown in table 4.1.

Table 4.1. The Monty Hall problem.

Major prize	You select	Monty reveals	You decide	Outcome
Door A	Door A	Door B or C	Stay	Win
			Switch	Lose
	Door B	Door C	Stay	Lose
			Switch	Win
	Door C	Door B	Stay	Lose
			Switch	Win
Door B	Door A	Door C	Stay	Lose
			Switch	Win
	Door B	Door A or C	Stay	Win
			Switch	Lose
	Door C	Door A	Stay	Lose
			Switch	Win
Door C	Door A	Door B	Stay	Lose
			Switch	Win
	Door B	Door A	Stay	Lose
			Switch	Win
	Door C	Door A or B	Stay	Win
			Switch	Lose
Total wins by strategy			Stay	3
			Switch	6

Looking at the table, one can see all of the possible combinations of winning door and selected door—including what this means for 'Monty's' action and whether switching or staying will receive the major prize. At the bottom of the table, one can see confirmation that switching is twice as likely to win as staying. (This can, and repeatedly has, also been confirmed via Bayes' theorem or via simulation.)

This is because Monty's action gives us additional information about where the prize is or is not, which changes our degree of uncertainty regarding the location of the major prize and allows us to update our probabilities. That this is not intuitive, however, was clearly demonstrated by the furore following Marilyn vos Savant's publication of the problem and solution, after which she received thousands of letters—including from mathematics professors—telling her she was wrong (Tierney 1991). That is, the Monty Hall problem neatly demonstrates both that probabilities are subjective and that people are not very good at intuiting the solution to probability problems.

4.2.2 Roll of the die

Perhaps a simpler demonstration of the subjective nature of probability is the following. Imagine that we are playing a game in which I roll a die and you and another person have to guess what number I have rolled. I have a standard, six-sided die marked with pips rather than numbers and am rolling it behind a screen.

On the first roll, I roll the die completely behind the screen. At this stage, your uncertainty— and that of the other player—regarding the outcome is informed only by what you know about the die: that it was a standard, unloaded die with six sides. That is, you should expect each of the six sides to be as likely as any other so your probability estimate for each number should be 1/6.

On the second roll, however, I roll it slightly too hard and it lands so that you— but not the other player—catch a glimpse of the die's top corner before I shift the screen to cover it; and you see that the corner is blank. That is, the top face does not have a pip in the corner that you saw. What does this mean for your uncertainty and the probabilities that you should assign to the six possible outcomes?

Thinking about dice, you realise that only three of the six faces do not have a pip in every corner—the 'one', 'two' and the 'three'. Therefore, these are the only possible values given the new information that you have. Additionally, the one-face has all four corners without pips while the others have only two. That is, you know you have seen a face without a corner pip and, of the eight corners of the die that do not have a pip, half are on the one-face and one quarter on each of the two- and three-faces. As a result, your glimpse allows you to update your probabilities to a 1/2 chance of my having rolled a one and a 1/4 chance (each) of my having rolled a two or three.

The other player, bin contrast, does not have this information and, as such, retains the original probabilities of 1/6 for each face. That is, the 'correct' probability is the one that matches your level of knowledge, not a function of the system.

4.3 Bias in intuitive probability theory

In addition to these problems, there are a number of known biases that occur when people are asked to solve probability problems that are relevant to scientists. Two of these—sample size invariance and base rate neglect—are discussed below.

4.3.1 Sample size invariance

Imagine that you have access to two robotic sorting machines designed to sex *Drosophila melanogaster*. The 15-Fly Machine takes a sample of 15 flies at a time while the 45-Fly Machine accepts a sample of 45 flies. Each then accurately counts the number of male and female flies in its sample and flags the sample as unbalanced if the proportion of male flies is 60% or above.

Over the course of a week's testing, each machine sexes 100 groups of flies. At the end of the week, which machine do you think will have flagged more samples as unbalanced? The 15-Fly Machine or the 45-Fly Machine? Or will the number be approximately the same?

The correct answer is that the 15-Fly Machine is likely to produce more than twice as many flagged samples and, for those used to thinking about sample size and statistical power, this may seem unsurprising. Given the assumed 50:50 sex ratio in the flies, the chances of getting 60% of a sample to be male depends on the size of the sample—corresponding to the binomial probability of obtaining 9 or more heads out of 15 coin tosses or approximately 30.4%. Getting 27 male flies in a sample of 45, however, is far less likely—approximately 11.6%. This is, simply, the law of large numbers—the tendency of larger samples to yield average values that are closer to the long-run expected value of a distribution.

That is, the larger the sample, the more closely we should expect the proportion of male and female flies to approximate the population average of 50% and, thus, the *less* likely it will be that the sample will be flagged as unbalanced by our hypothetical machines.

The concern for science and its communication, however, can be seen in people's responses to problems like these. In Tversky and Kahneman's (1974) statement of this problem, 21% of respondents correctly selected the smaller sample as being more likely to deviate from the expected value, 21% the larger and the remainder indicating that the two samples were (approximately) equally likely to do so. Somewhat surprisingly, this holds true when testing people with science and engineering backgrounds (Welsh *et al* 2016). Across 10 years of testing oil industry samples consisting primarily of engineers and geoscientists, only 23% of respondents correctly identified the smaller sample as being more likely to deviate.

Of course, when this is explained to people with a background in probability theory, they invariably recognise their error and agree that the smaller sample is the correct answer. The worry is that this is, despite such knowledge, not their initial response. Even more worryingly, the rate at which people select the correct option is *less* than chance performance on a three-choice question (i.e. 33%); that is, were people just guessing, we would expect them to do better than they are observed to do. So, it is not just that people do not have any idea but that their intuitions are actively leading them astray—resulting in them regarding the two samples as being equally reliable sources of information about the underlying distribution.

This, of course, has dramatic implications for the communication of science. Specifically, people's invariance to sample size means that most readers of a scientific paper or missive will disregard the sample size when interpreting the findings, with the result that they may markedly over-estimate the reliability and thus replicability of the findings—a theme that will be expanded upon in the discussion of publication biases in chapter 11.

In fact, in general, people are convinced of a finding's reliability not by its sample size but rather by a variety of other factors such as their trust in the source of the information (Welsh and Navarro 2012) and the strength and cogency of the proposed causal mechanisms (see, e.g. Krynski and Tenenbaum 2007).

4.3.2 Base rate neglect

Such effects also affect another commonly studied area of intuitive statistics—base rate neglect, which describes the tendency for people to ignore or underweight information about the prior rate of occurrence (base rate) of some event once presented with additional, diagnostic or descriptive information. While there are a number of variants designed to show base rate neglect under different conditions, a fairly typical base rate question (based on Cascells *et al* 1978) runs as follows:

Imagine that you are a doctor engaged in a testing program for BRN syndrome (BRNS). In the population at large, 1 in 1000 people (0.1%) are known to have a genetic mutation causing BRNS. Your test is 100% accurate in detecting this mutation when the patient actually has it but also has a 5% false positive rate. You have just tested a randomly selected person and their test has come back positive for BRNS. In the absence of any other information, what is the probability that the patient has BRNS?

The solution to this problem depends on understanding how to update probabilities in light of this, new information. What we have, in terms of the relationship between the test and the actual occurrence of the syndrome, is P(+ve | D)—the probability of a positive test given the disease being present; that is, the test's accuracy. What we are being asked for, however, is P(D | +ve)—the probability of the person having the disease given that the test is positive.

Most people, however, fail to draw this distinction and give an answer that is very close to the accuracy of the test. In Cascells *et al*'s (1978) data, for instance, the most common response was 95% and, across a range of experiments on base rate neglect (see, e.g. Bar-Hillel 1980, Welsh and Begg 2016, Welsh and Navarro 2012) this pattern is repeated, with people—including doctors and scientists—selecting the accuracy rate of the test or adjusting from it slightly to produce their estimate. As can be seen in the problem above, 95% corresponds to the difference between the accuracy and false positive rates of the test—reflecting people's recognition that they need to combine the probabilities but their lack of knowledge as to how this needs to be done. The selected combination (accuracy minus false positive) also reflects the common finding after which the effect is named—neglect of the base rate information.

In fact, to solve the problem requires all three pieces of information (base rate, accuracy and false positive rate) to be combined via Bayes theorem or equivalent use of probability theory. Perhaps the clearest way of showing this, however, is via a probability table such as table 4.2.

In the base rate problem, we are told that the test returns a positive result and what we are asked to determine is how likely it is that this positive result actually indicates the presence of the syndrome—that is, how often a positive test is a true positive. Looking at table 4.2, one sees the four possible states of nature resulting from the two binary conditions: the person having BRNS or not; and the test being

Table 4.2. Probabilities in the base rate neglect problem.

Base rate		Test result		Probability	P (BRNS | +ve test)
BRNS	0.1%	Positive (True)	100%	0.1%[a]	= True +ve / +ve
		Negative (False)	0%	0%	= 0.1/(0.1+4.995)
					= 0.1/5.095
No BRNS	99.9%	Positive (False)	5%	4.995%[a]	= 0.0196 (1.96%)
		Negative (True)	95%	94.905%	

[a] Two ways in which a positive test can result.

positive or negative. The associated probabilities of these outcomes are derived from the base rate and either the accuracy rate (where the person actually has BRNS) and the false positive rate (where they do not). This yields two ways in which a positive test can result—a true positive and a false positive (the highlighted portions of the table).

Calculating the probability we need is then as simple as dividing the true positive rate by the total chance of a positive result (i.e. true positive plus false positive), which yields a less than 2% chance of the positive test actually reflecting a true positive—because the base rate is so low, false positives are far more likely than true. The difference between this and the typical response of returning an estimate close to the accuracy of the test is stark—a factor of more than 40 in this case.

People's inability to intuitively understand how probabilities should be combined therefore can have severe implications for those interested in the clear and accurate communication of information and for the decisions based on people's probability assessments. For instance, a doctor who believes that the positive test actually means that there is a 95–100% chance the patient has a disease is likely to move immediately to treatment. This is, of course, unnecessary for 98% of patients, wasting resources and potentially causing harm as most treatments have adverse side-effects. If, in contrast, the doctor recognised that the positive test really only indicates a 2% chance of the disease being present, it is far more likely that they would recommend further testing in an attempt to confirm or refute the initial test avoiding this waste and injury.

The observation regarding how few people can meaningfully incorporate these probabilities also has implications for how we should communicate information about test accuracy and false positive rates to others. For example, we know that telling a patient the accuracy rate of a test is most likely to fix that in their minds as the probability of their having whatever the test is testing for—regardless of whether base rate information or false positive information is given simultaneously. Relying on people to know Bayes theorem and understand that they should use it in this circumstance is a flawed plan.

Instead, those interested in accurately conveying probability information should ensure that the information is presented in a way that makes the necessary comparisons as clear as possible. Using a probability table such as table 4.2, for example, greatly increases the proportion of people who accurately incorporate the different sources of information (Sloman *et al* 2003) as it clarifies which probabilities need to be compared. A number of researchers also argue that presenting data in terms of natural frequencies (e.g. 1 in 1000) rather than percentages assists people in avoiding base rate neglect and thus improves probability updating, despite the mathematical equivalency of the two (see, e.g. Cosmides and Tooby 1996)—a theme that is explored in greater detail in chapter 6.

4.4 Conclusions

The examples in this chapter barely scratch the surface of the literature on human intuitions about probabilities but are, I hope, sufficient to make clear that the quotation from Laplace presented in chapter 3 is overly optimistic. It is not the case that people intuitively understand probability theory. In fact, people's intuitions about how to interpret and combine probabilities often lead them well astray of good reasoning and result in estimates that are wildly divergent from the actual probability of events given by the application of probability theory.

Importantly, this holds true for well-educated, scientifically literate people. That is, simply knowing probability theory seems insufficient to guard against the biases discussed above (and similar errors in reasoning). Rather, care needs to be taken to structure information in ways that are more intuitively accessible to people so that we are not led astray in our understanding and can clearly communicate probabilities to others.

Finally, people's failure to understand or apply probability theory has consequences for the communication of almost all scientific observations. This is because our understanding and interpretation of statistical inference is underpinned by our knowledge of probability theory. So, the fact that people do not intuitively understand probability—including core aspects such as the law of large numbers—means that they are very likely to misinterpret statistical evidence. As communicators in possession of this knowledge, it becomes our responsibility to make every effort to ensure that statistics are re-interpreted in such a way as to allow people to understand both the evidence that has been found and how reliable that evidence should be taken to be.

References

Bar-Hillel M 1980 The base-rate fallacy in probability judgments *Acta Psychol.* **44** 211–33

Bratvold R and Begg S 2010 *Making Good Decisions* (Richmond, TX: Society of Petroleum Engineers) ch 1

Begg S H, Bratvold R B and Welsh M B 2014 Uncertainty versus variability: what's the difference and why is it important? *Society of Petroleum Engineers Hydrocarbon Economics and Evaluation Symp. (May)*

Casscells W, Schoenberger A and Graboys T B 1978 Interpretation by physicians of clinical laboratory results *New Engl. J. Med.* **299** 999–1001

Cosmides L and Tooby J 1996 Are humans good intuitive statisticians after all? Rethinking some conclusions from the literature on judgment under uncertainty *Cognition* **58** 1–73

Knight F H 1921 *Risk, Uncertainty, and Profit* (Boston, MA: Hart, Schaffner and Marx; Houghton Mifflin)

Krynski T R and Tenenbaum J B 2007 The role of causality in judgment under uncertainty *J. Exper. Psychol. Gen.* **136** 430

NASA 2016 Global mean estimates based on land and ocean data, Goddard Institute for Space Studies https://data.giss.nasa.gov/gistemp/graphs/ (Accessed: 31 August 2018)

Selvin S 1975 On the Monty Hall problem (letter to the editor) *Am. Statistician* **29** 134

Sloman S A, Over D, Slovak L and Stibel J M 2003 Frequency illusions and other fallacies *Organ. Behav. Hum. Decis. Process* **91** 296–309

Tierney J 1991 Behind Monty Hall's Doors: Puzzle, Debate and Answer? *The New York Times* 21 July www.nytimes.com/1991/07/21/us/behind-monty-hall-s-doors-puzzle-debate-and-answer.html (Accessed: 1 September 2017)

Tversky A and Kahneman D 1974 Judgment under uncertainty: heuristics and biases *Science* **185** 1124–31

Welsh M B and Begg S H 2016 What have we learned? Insights from a decade of bias research *APPEA J* **56** 435–50

Welsh M B and Navarro D J 2012 Seeing is believing: priors, trust, and base rate neglect *Organ. Behav. Hum. Decis. Process* **119** 1–14

IOP Publishing

Bias in Science and Communication
A field guide
Matthew Welsh

Chapter 5

Truth seeking? Biases in search strategies

This chapter's focus is on the way in which people search for new information and update their beliefs as a result. This is, of course, a central aspect of all aspects of science and communication: we are constantly coming into possession of new information that might alter what we believe or make clear to us that we simply do not know enough about some event or object. When this happens, we need to decide what to do about this sudden challenge to our beliefs. That is, do we need to search for new information? If so, how and for how long need we search become important questions.

The ways in which we choose to do this, however, are rarely as rational as we might hope. Instead, they are influenced by the same constraints as our other forms of decision making. Specifically, we tend to rely on simple, heuristic search processes and, as a result, systematic biases are observed in our decisions and thus knowledge.

5.1 Bounded rationality in search

A key point to recall here is that the *Homo economicus* model of human behaviour relies on the tenets of rational decision making (von Neumann and Morgenstern 1945). That is, to act in a completely rational manner, a person needs to meet these criteria. These criteria, however, are defined as relationships that must hold amongst the complete set of options. For example, the axiom of independence holds that a person's preference between two options, A and B, should be independent of any other outcome or option. That is, if you prefer A to B, you should do so whether you are selecting from amongst the set of A, B and C or A, B and D. In a gambling task or simple, risky-choice decision, this is easily assessed as all of the options are defined in advance. In real-world decisions, however simply determining what the options are becomes a decision task in and of itself and it becomes significantly harder to demonstrate whether such rationality exists.

Gigerenzer and Goldstein (1996) argue that real-world decision making leads to an exponential expansion of the difficulty of making a rational decision—referring to the assumptions required for rationality as requiring the mental abilities of a 'Laplacian demon'. To make clear why this is, consider what seems like a relatively simple decision—purchasing a new work computer. When doing so, we will, undoubtedly, have some criteria in mind—the tasks we need to use it for, the programmes we need it to run, how urgently it is needed and so on—but, taking these as given for now, let us just think about the options that are available.

That is, how many computers are there that might meet your criteria and which a fully rational person should, therefore, include in their decision process? The answer, of course, depends on how stringent your criteria are but, even with a set of criteria in place (cut-offs for price or processor speed or other key indicators) the set of available options is almost unknowably large—particularly when design-your-own or build-your-own options are included.

You can, of course, use any one of your criteria—or even a set of criteria—to limit your search but such an approach does run the risk of, for example, eliminating a computer that is slightly above your preferred price but which is, in every other way, markedly superior to all other available options. That is, using a single criterion to limit a search runs the risk of selecting from amongst a set of options that has already excluded the option that would be judged optimal when looking at all criteria simultaneously—which fails to meet the rationality criteria. Even using multiple criteria as simple cut-offs will fail to meet a standard of rationality—as multi-objective decision making requires that objectives be weighted against each other simultaneously.

In addition to this need to assess an option on all criteria simultaneously, we need to consider how much time and effort we can afford to put into the search process. This will be affected by how urgently the new computer is needed but also by how much you value your time as time spent searching for computers is time that cannot be spent undertaking other tasks. The result of this is that you almost certainly will not have the time to find, let alone consider and score, all of the available computers.

As a result, the above process will not meet the criteria for rational decision making and, as Gigerenzer and Goldstein note, the majority of situations involving search for options or information in the real world seem destined to run into similar problems. This, of course, implies that people's search strategies must be non-rational, which will tend to lead to predictable, systematic biases.

5.1.1 Selecting a secretary

In addition to the complexity of searching for and through alternatives as described above, the world imposes additional constraints on our decision making in the form of competition for resources. To continue the example above, for instance, it might be that there is a limited number of certain types of computer available, with the result that spending too much time on the search will result in some of the options becoming unavailable. This added pressure necessitates a different search strategy, one that is often discussed in terms of a traditional decision-making task called the 'secretary problem' (see, e.g. Seale and Rapoport 1997).

This problem (also called the 'marriage', 'dowry' or 'googol' problem) describes an optimal stopping problem (in a somewhat archaic fashion). Specifically, the idea is that a manager is attempting to select the best secretary from a pool of candidates. These candidates are, however, sent from the secretarial pool one-by-one and the manager can either select to hire them or send them back to the pool. If a secretary is not hired, however, they cannot be called back at a later stage as they have, presumably, been sent out to and selected by someone else. That is, it reflects selection tasks where returning to previous options is problematic (thus, the 'marriage' problem where it is framed as selecting a partner, in line with the common, Western expectation that we will only entertain/date one serious partner at a time) or, more generally, any situation where previously seen options are likely to be taken by others, as might be the case for limited edition or on-sale items.

Variants of this problem include whether the number of candidates is known in advance and whether the scoring system used to assess candidates gives them a score on a scale or simply ranks them against previously seen candidates (Ferguson 1989). As one might expect, knowing the number of candidates in advance makes the problem more easily soluble—as does information beyond the rank of the current applicant (e.g. if you can score candidates on a 0–100 scale, then seeing a candidate who has scored 99 is far more informative than simply seeing that that candidate is ranked 1st). Even in the simplest cases, however, the determination of the optimal (rational) rule for stopping the search is a mathematical task beyond most people's capabilities.

Experimental work, however, has demonstrated that, while people's performance on these sorts of tasks is sub-optimal (Seale and Rapoport 1997), it can be quite close to that benchmark (Lee et al 2004). That is, despite the fact that people are incapable of solving the problem rationally, many of them manage to solve it in a manner that produces results almost as good.

Taking the (conceptually) simplest version of the problem—a known number of ranked candidates—as an example, it becomes clear why this could be. In this formulation, as a candidate appears, you immediately know whether they are the best candidate or not and should only accept them if they are. The only situation where you do not select the highest ranked candidate of those you have seen is where you have come to the last candidate without making a decision and are, as a result, stuck with them regardless of their rank.

So, the problem facing the decision maker here is to work out how to predict whether a later candidate will be better than the current number 1 ranked candidate. Obviously, this is a function of the number of candidates seen and yet to be seen, as shown in table 5.1.

Looking at the table, one sees a visual representation of a secretary problem with ten candidates who are assessed in terms of their rank. The *a priori* probability of any particular candidate being the best overall of the ten candidates is, of course 0.1. The black face, however, indicates the current candidate and, as such, the *only one* that can be selected at that point in the task. Looking at the probabilities, one sees that, as the task proceeds, the probability of the current candidate being the highest ranked of those seen so far decreases— from 1 for the first candidate (because no one

Table 5.1. The secretary problem example.

	Candidates										Probabilities		
											Overall rank =1	Current rank =1	Better to come
Trial	1	2	3	4	5	6	7	8	9	10			
1	☻										0.1	1	0.9
2	☺	☻									0.1	0.5	0.8
3	☺	☺	☻								0.1	0.33	0.7
4	☺	☺	☺	☻							0.1	0.25	0.6
5	☺	☺	☺	☺	☻						0.1	0.2	0.5
6	☺	☺	☺	☺	☺	☻					0.1	0.17	0.4
7	☺	☺	☺	☺	☺	☺	☻				0.1	0.14	0.3
8	☺	☺	☺	☺	☺	☺	☺	☻			0.1	0.125	0.2
9	☺	☺	☺	☺	☺	☺	☺	☺	☻		0.1	0.11	0.1
10	☺	☺	☺	☺	☺	☺	☺	☺	☺	☻	0.1	0.10	0

else has been seen yet) down to 0.1 for the last one. Similarly, however, the probability of seeing someone better also decreases as more options are seen.

This means that the decision maker needs to trade off the fact that people seen earlier are more likely to be highly ranked—and thus, in principle, be selectable—with the fact that early selections also increase the likelihood that a better candidate is yet to come. That is, the goal is to make a selection early enough to avoid being put in a position where you are left with the last candidate but late enough that the person you select has a reasonable chance of being the best candidate overall. Mathematically, the solution to this problem is a stopping rule of the type: disregard the first $r–1$ candidates and then select the first person thereafter who is ranked '1'. Using this rule will select the best candidate ~37% of the time from amongst a large group of candidates (and somewhat more often in smaller groups).

Seale and Rapoport's (1997) study suggested that people do exactly that—with a slight bias towards lower values of r than the optimal solution suggests—with the result that human performance is somewhat sub-optimal, but still far better than chance.

5.1.1.1 Satisficing
This sort of decision thus has clear connections to the work of Herbert Simon (1956), who introduced the idea of satisficing as a description of how people make decisions in environments where optimisation is too difficult a task—whether because of our limited cognitive processes or due to the structure of information in the environment (e.g. spatio-temporal distances between options requiring that they be considered as a sequence rather than in parallel). Specifically, the idea is that, rather than attempting the very difficult (or impossible) mathematics required for making optimal decisions, decision that are good enough can be made using simple, heuristic rules.

For example, in the secretary problem, setting a cut-off rule such as viewing one-third or one-half of the candidates before making a decision is a cognitively undemanding decision rule but one which is easily formulated by people and which comes very close to replicating the optimal mathematical solution. Thus, given constraints on people's time and available effort, such a heuristic is, arguably, more 'optimal' than attempting to calculate the optimal solution.

Satisficing, in its specific sense, refers to the selection of the first option that meets a pre-determined benchmark. For example, when buying petrol, a person will probably not engage in any form of rational assessment of how and when to buy their petrol—rather they will have a benchmark price in mind and, as soon as they see petrol at or below that price, they will stop and buy it. That is, it reflects a serial search for a satisfactory option.

5.1.2 Naturally naturalistic

The type of search described above comes naturally to people and the part of the reason for this seems to be that it actually produces *better* outcomes than attempts to be rational in a variety of environments. A classic example of this comes from the naturalistic decision making (NDM) literature (see, e.g. Klein 2008 and, for more detail, Klein and Zsambok 1997), which focuses its attention on how expert decision makers operate in real-world environments.

NDM has included work with firefighters and military commanders and a key takeaway from this research is that the combination of time pressure and high consequences for decisions results in such personnel relying on satisficing as their primary search strategy. That is, despite the researchers' initial belief that the expert decision makers would rely on some sort of rational (if constrained) optimisation—comparing multiple strategies and selecting the one best suited for the current problem—the process that the personnel described involved no comparisons at all. Instead, on examining a situation, they were immediately struck with a possible solution and then immediately implemented it—so long as it seemed satisfactory. In fact, some denied ever having *made* a decision, relying on intuitions operating far beneath the conscious level.

While initially surprising to NDM researchers, the evidence clearly supports what has come to be the standard NDM view—that expert decisions makers in these high-consequence environments satisfice rather than attempting to optimise. The reason that this works is that the time pressures involved are such that any delay to attempt to use rational processing costs more than the rational process might gain. Imagine, for instance, coming home to find a small fire in your kitchen. Should you stop and engage in a rational process coming up with multiple ways of fighting the fire and then deciding between them? Or should you simply try to put it out using whatever you know you have in the kitchen? Clearly, in such situations, any time used considering options results in the fire becoming harder to contain and thus decreasing the likelihood of a good outcome.

Now, of course, in the heat of the moment and not being an expert firefighter, you could make a mistake—throwing water on an electrical or grease fire, for instance.

This is the difference between an expert and a novice in these situations: an expert has the knowledge and experience to know how best to fight these fires. This knowledge is built up over years of training and experience and is called, by NDM researchers 'situational awareness'. The important thing to note, however, is that this knowledge has become automatic—operating beneath consciousness—with the result that the expert decision maker simply looks at the situation and their situational awareness results in a single, plausible course of action intuitively coming to them, which they immediately implement so long as it seems satisfactory.

The implications of these NDM findings for decision making more generally are worth considering. First, many of our most important decisions are made under time pressure, making fast, heuristic-based search for strategies or information the better option. Second, expertise does not always make an expert aware of their own situational awareness, rather leading to them having strong but sub-conscious intuitions about what they should do. When we add this to the fact that most of our everyday decisions (what to wear, what to have for lunch, etc.) are not valuable enough for us to invest the time and effort of rational decision making, we are left in a situation where the vast majority of the decisions that we make are made in this fast, heuristically based manner. That is, naturalistic decision making is natural to us and we feel that satisficing is perfectly satisfactory. In fact, decisions being made quickly and confidently is something that people commonly admire.

However, given the society in which we live, you can mount a strong argument for us now being in a situation where we are making very important decisions but where we have the time and processing power available to actually engage in some sort of optimisation process. For instance, policy decisions about budget allocations and research directions or personal decisions about retirement funds or even career change. In these cases, people's natural tendency will be towards making decisions in the same way that they typically do—quickly settling on a satisfactory option. Given more time and effort put into the search for options, however, significantly better outcomes might be achieved.

5.2 Recognising recognition

The above sections discuss how people tend to search for options and suggests that people will stop as soon as they have found one that is satisfactory for their current decision. This is, however, only one sort of decision. In other decisions we already have multiple options and are being asked to judge or select which better meets our criteria. Returning to our buying a computer example from section 5.1, above, we can imagine having search the internet with our basic criteria. We are now staring a screen full of options that meet our set of specifications. How should we select amongst them?

As noted above, we have a variety of competing objectives—but most of these have already been dealt with by our search parameters. What we are left with is a set of options, all of which seem satisfactory, and what we want is to choose the best one from amongst them. How then, do we proceed?

The answer is that most people, all other things being equal, will tend to select an option that they are familiar with—a brand or model of computer that they recognise. 'Why is this?' and 'is it a good strategy?' are questions that will be answered after your...

5.2.1 German geography test

Look at the following list of twelve German state capitals and consider: which of these are the largest cities?

Berlin	Hanover	Saarbrucken
Bremem	Mainz	Schwerin
Dusseldorf	Munich	Stuttgart
Hamburg	Potsdam	Wiesbaden

While accurately ranking them from largest to smallest may be beyond most people, when I present these cities to (non-German) students in pairs and ask which of each pair is the larger they, overwhelmingly, select the correct city. That is, despite not actually knowing very much about German geography, they are able to correctly determine which option meets the criterion of larger size. The way in which they do so is to simply select the option that they are more familiar with or—in the most extreme cases—the only one of the pair that they recognise.

5.2.2 Recognition and other selection heuristics

This 'recognition heuristic' is a surprisingly effective decision rule, with Goldstein and Gigerenzer (2002) presenting research showing that recognising one city from a country (and not the other) is an extremely good predictor of the cities' relative sizes—accurate in around 90% of cases. This, does, however, lead to a very strange result—that knowing more can actually reduce your accuracy.

Specifically, in cases where you recognise both options, you cannot rely on recognition to determine which of the two cities is larger and this leads to the perverse outcome that students in the US (who know more US geography) do worse, at 62% accuracy, in selecting the larger of two US cities than do German students (Goldstein 1997, cited in Gigerenzer and Goldstein 2011) who are more likely to *only* recognise the larger and achieve 90% accuracy. The US students are left in the situation of having to look for other discriminating cues that may not be as predictive (e.g. the take-the-best heuristic, which describes a process undertaken when simple recognition fails to distinguish between options—in effect, looking down a list of potentially distinguishing cues until the first one is found that actually splits the options, see Gigerenzer and Goldstein 2011).

So, how is this relevant to our selection of a computer? Well, it turns out that the recognition heuristic is effective across a variety of environments. For instance, recognition of a company name is predictive of that company's performance. That is, imagine you are selecting which of two technology companies to invest in. If you recognise one and not the other, chances are that the one you recognise is the larger and more profitable company and so that is probably the safer investment.

Similarly, in our thought experiment, the computer brands you recognise are most likely to come from larger companies and, in the absence of additional information, it is not unreasonable to expect that their products will be superior—in reliability or performance. While not a perfect relationship, the underlying assumption here is that these companies have become successful *because* their products are good and thus the recognition of the brand can be an indicator of quality.

So, the recognition heuristic works in a variety of contexts, allowing us to quickly discard unfamiliar options and thus simplify our decisions. Even where it does not work, however, we tend not to revert, naturally, to rational decision making. Instead, there are additional heuristic processes that we engage in—such as take-the-best (described above) or the fluency heuristic, where we judge which alternative is better based on how *quickly* it is recognised (Schooler and Hertwig 2005).

These heuristic decision processes underlie a wide range of our common decisions and work surprisingly well using just the information that we already have to hand. That is, they require no search for additional information beyond our own memories and produce outcomes that are good enough for most people's decisions. Indeed, the results presented by Gigerenzer and colleagues suggest that searching for additional information can actually undermine performance. The danger, here, is that—as in the satisficing and naturalistic decision-making processes described above—our natural tendencies seem to be towards limiting the search for additional information.

That is, people's natural tendency is to, first of all, not search for additional information or options—just selecting from amongst those already available—and then, if a search is necessary, to terminate that search as soon as possible.

The question that needs to be asked, however, is whether this is actually important? Given the above results, it seems that this heuristic processing leads to good, if not necessarily optimal, outcomes. Should we, therefore, be happy sticking with this as our decision-making process? Some researchers, including Gigerenzer, argue that (to some extent, at least) we should; that the heuristic processes lead to outcomes that are often very good and rational processes are just too hard to implement meaningfully (the Laplacian demon argument stated above).

A central observation coming from research into heuristics, however, is that they are not general purposes tools. This is important for two reasons. First, we do not have one or two heuristics that we use for all tasks. Rather, as noted in the aside on two systems theories at the end of chapter 2, a wide variety of unconscious, heuristic processes used for different tasks comprise system 1.

Second, the basic idea of bounded rationality is that heuristics work in specific environments—the two blades of Simon's scissors. That is, a heuristic can only be expected to produce good outcomes in a limited set of environments—where the process and the information structure match.

By way of example, consider the sets of cities in table 5.2 below. We have 24 cities in total— four from each of three US states and three Australian states. The question is: which city, from within each set of four, is the state capital?

Assuming you do not already know the answer, how do you then go about solving it? One way is via the recognition or fluency heuristics described above—that is, whichever city is recognised (or recognised most quickly) should be selected as the

Table 5.2. State capitals.

USA			Australia		
New York	California	Ohio	NSW	Victoria	South Australia
Albany	Los Angeles	Akron	Albury	Ballarat	Adelaide
Buffalo	Sacramento	Cleveland	Newcastle	Bendigo	Mount Gambier
New York	San Diego	Cincinnati	Sydney	Geelong	Murray Bridge
Rochester	San Francisco	Columbus	Wollongong	Melbourne	Whyalla

capital. This makes sense as one can assume that capital cities are more likely to be recognisable—more likely to be mentioned in the media and so forth. In this instance, however, such an approach offers a mixed performance.

For the Australian cities, the most recognisable in each list is, indeed, the capital (i.e. Sydney, Melbourne and Adelaide). For the US cities, however, this does not hold true—the most easily recognisable cities (New York, Los Angeles and Cleveland) are not the capitals. This seems to come down to people's beliefs about the attributes of capital cities, as learnt from their particular environment. In Australia, capital cities are, without exception, also the largest cities in their respective states and this learnt association of 'capital = large' accords with many other examples from around the world. For example, more than 80% of countries have their largest city as their capital.

In the US, however, the relationship is much weaker with only 16 of 50 states having their largest city as their capital. This results in many non-Americans, at least, making errors on this task because they apply the rule they have learnt in one environment to another where the environmental information structure is quite different. That is, while people still tend to use recognition to guide their judgements, the validity of recognition as a cue for capital city status varies across environments.

More generally, consider the question—what environments are our cognitive processes adapted to work in? Are these the same environments that we now operate in? While the above example focuses on learnt environmental cues, it is also worth considering how deeply ingrained our cognitive processes might be. Taking an evolutionary view-point, it seems probable that the rate of change of our cognitive processes is likely to lag behind the societal changes that have occurred over the past few millennia, since the industrial revolution and over the last 50 years in particular. For instance, are the cognitive processes required to function appropriately in the sort of world occupied by nomadic or agrarian tribal societies the same ones as required to operate in a global, technological society? To the extent that you believe the structure of the environment no longer matches the heuristic processes available, you must expect the heuristics to give biased results and therefore, there are reasons to distrust them.

Take, for example, the central example of this chapter—the search for options. Assuming a nomadic lifestyle, limiting the time and effort spent on the search for options may be an excellent decision strategy as continued searching takes substantial time and will result in closing off other options (e.g. travelling west to search for resources rules out any possibility of finding resources that lie to the east). Additionally, deciding not to exploit resources at your current location in the hope of better resources further on is a high-risk strategy as those future resources are uncertain.

In contrast, in a global, technological society, the search for options often requires far less time and effort. While we may not enjoy trawling through internet sites for exactly what we want, we can do so without any need to travel and without limiting the parts of the world that we are actually searching. In most cases, this means that time spent on search also has limited impact on the availability of options. That is, we can bookmark a good deal on a computer and continue searching without fear of the effects of scarcity—excepting in the case where there is a time-limited deal, it is a safe assumption that there will continue to be more computers if you continue searching and that you can return to a previous 'location' later to still find the same option there.

That is, the environment that we are operating in now bears little resemblance to the one in which our search heuristics likely evolved. As a result, continuing to rely on these heuristics is likely to result in our spending less time and effort searching than we really should in order to make better decisions and achieve our goals.

5.3 Search strategies

The sections above argue that people tend not to spend enough time and effort looking for options and information, leading them to terminate their searches before they get a good idea of the range or extent of information or options that are actually available. This is problematic for people interested in making good decisions or relying on people to, for example, educate themselves about some topic. More troubling, however, is the way in which searches are often conducted. Look, for example, at figure 5.1.

Figure 5.1 shows a scenario in which a new, computerised test has been proposed as a replacement for an older, manual one. The key claim of the new test is that it

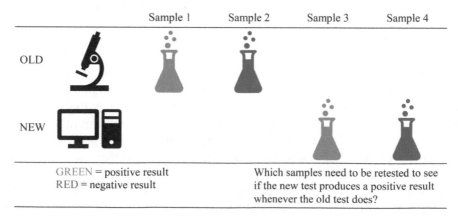

Figure 5.1. Old test versus new test.

5-10

can serve as a replacement because it produces a positive result in any case that the old test does. Currently, four samples have been tested, two with each test—getting one negative and one positive result in each case. The question is: which of the four samples need to be retested in order to test whether the new test is performing as advertised? Obviously, only *required* tests should be conducted so as not to waste time and effort.

5.3.1 Confirmation bias

This task is a variant on the Wason selection task (Wason 1966) and the observation from many years of research on such tasks is that people show a clear bias in how they choose to search amongst the options. Logically, to test the claim that the new test will produce a positive result any time the old test does, one need only retest two samples—1 and 4. The reason for this is that sample 1 currently has a positive result from the old test and so a negative test here from the new test will demonstrate the claim to be false. Similarly, sample 4 currently has a negative test for the new test and so a positive result from the old test will demonstrate the claim to be false. Retesting the other samples does not help us with the claim. Sample 2 has a negative result from the old test but no claim has been made about what the new test will do when the old test was negative, so regardless of the result of the new test, this is consistent with the claim. The same goes for the positive result from the new test on sample 3—either a positive or negative result from the old test is consistent with the claim.

This is not, however, how most people answer the question. Instead, the typical response is to select samples 1 and 3—that is, the options that offer the possibility of *confirming* rather than testing the rule. This tendency is given as an example of the confirmation bias which, while commonly demonstrated with these simple, logical examples, occurs in a broad range of situations and has important, real-world implications for how people search for information (see, e.g. Nickerson 1998).

In general, the observation is that people search for evidence that accords with or confirms their current hypothesis—even when, as is the case in the above example, a search for disconfirming evidence would be a better strategy (note, however, that this is not always the case; in some instances a positive search strategy is more informative, as outlined by Klayman and Ha 1987). Additionally, evidence that supports an already held position is viewed more favourably. The result of this is, unsurprisingly, the tendency for people to reinforce things that they already believe. The 'echo chamber' effect of the internet is often argued to result from this, with people easily able to find 'evidence' that supports their opinion, which can contribute —as a result of the ease with which they find such evidence—to them feeling that a majority of people must agree and that those who do not are outliers. That is, the false consensus effect (Ross *et al* 1977).

Despite the explicit focus of science on falsifiable hypotheses, biases like these affect scientists' judgements just as they do lay-people. For instance, scientists and science students have been shown to display confirmation bias, judging evidence that aligns with their own previously stated theories as more convincing than evidence to the contrary (Koehler 1993).

5.4 Conclusions

The above argues strongly for a need to understand how people engage in search for new information and options—which are key aspects of both the practice and communication of science. To some extent, given these tendencies and the growth of online information, an increase in miscommunication or misunderstanding of scientific results or regarding the existence (or otherwise) of a scientific consensus may be unavoidable. Until online search engines come with integrated fact-checking, people's natural search strategies are likely to lead them to search for information that supports what they already believe, discount evidence that counters it and overestimate the proportion of the population who hold minority views. Where this will be relevant, however, is when a scientist wants to directly communicate information to others. Even here, however, understanding how people incorporate new information is important. The observation that people tend to weigh contrary evidence less heavily than confirmatory evidence, for example, plays a significant role in the difficulties scientists face in debunking factoids and other false beliefs—a topic that will be discussed in depth in chapter 13.

Where this knowledge has the potential to make a larger impact, however, is amongst scientists who would (we hope) hold the objective truth to be more important than their current opinions and thus be more willing to change their behaviour if they realise they have an unconscious tendency towards biased search strategies. For example, recognising that people (scientists included) have a natural tendency to underweight data that do not accord with their hypotheses may prompt more scientists to examine counter-intuitive or unpredicted results more closely. This, in turn, could help to reduce publication bias, as we will discuss in chapter 11.

References

Ferguson T S 1989 Who solved the secretary problem? *Statistic. Sci.* **4** 282–9

Gigerenzer G and Goldstein D G 1996 Reasoning the fast and frugal way: models of bounded rationality *Psychol. Rev.* **103** 650

Gigerenzer G and Goldstein D G 2011 The recognition heuristic: a decade of research *Judgm. Decis. Making* **6** 100

Goldstein D G and Gigerenzer G 2002 Models of ecological rationality: the recognition heuristic *Psychol. Rev.* **109** 75

Klayman J and Ha Y W 1987 Confirmation, disconfirmation, and information in hypothesis testing *Psychol. Rev.* **94** 211

Klein G 2008 Naturalistic decision making *Hum. Factors* **50** 456–60

Klein G A and Zsambok C E (ed) 1997 *Naturalistic Decision Making* (Mahwah, NJ: Laurence Erlbaum)

Koehler J J 1993 The influence of prior beliefs on scientific judgments of evidence quality *Organ. Behav. Hum. Decis. Process* **56** 28–55

Lee M D, O'Connor T A and Welsh M B 2004 Decision-making on the full-information secretary problem *Proc. 26th Meeting of the Cognitive Science Society* ed K Forbus, D Gentner and T Regier (Austin, TX: Cognitive Science Society) pp 819–24

Nickerson R S 1998 Confirmation bias: a ubiquitous phenomenon in many guises *Rev. Gen. Psychol.* **2** 175

Ross L, Greene D and House P 1977 The 'false consensus effect': an egocentric bias in social perception and attribution processes *J. Exp. Soc. Psychol.* **13** 279–301

Schooler L J and Hertwig R 2005 How forgetting aids heuristic inference *Psychol. Rev.* **112** 610

Seale D A and Rapoport A 1997 Sequential decision making with relative ranks: an experimental investigation of the 'secretary problem' *Organ. Behav. Hum. Decis. Process* **69** 221–36

Simon H A 1956 Rational choice and the structure of the environment *Psychol. Rev.* **63** 129

von Neumann J and Morgenstern O 1945 *Theory of Games and Economic Behavior* (Princeton, NJ: Princeton University Press)

Wason P C 1966 *Reasoning New Horizons in Psychology* ed B Foss (Harmondsworth: Penguin)

IOP Publishing

Bias in Science and Communication
A field guide
Matthew Welsh

Chapter 6

Same but different: unexpected effects of format changes

In this chapter, we will examine how the way in which information is presented to people affects their interpretation of that information. Specifically, how changing the format in which data are presented or other, surface, features can result in people putting more or less weight on the information that the data convey—despite the logical equivalence of the different presentations.

This results in the curious situation where the decisions that people make are shaped—in part at least—by the way that others have chosen to convey information to them rather than any intrinsic preference of the decision maker. That is, many of the decisions that we feel we make are, in some sense, being made for us by the people who decide how the information we receive will be formatted. Understanding how and why these effects occur is, thus fundamental to decisions about how we should and should not communicate scientific findings.

6.1 Percentages and frequencies

Which of the following $10 tickets would you prefer to buy? Ticket 1 gives you a 1% chance of winning a $1000 prize. Ticket 2 gives you a 1 in 100 chance of winning a $1000 prize. Most people, given a choice like this, would probably be a little confused. The two options are, clearly, identical and so there is no reason to prefer one over the other.

Imagine, though, that instead of a single person assessing them side-by-side, two groups of people were presented with these options—one group being offered the chance to buy ticket 1 (the % ticket) and the other group being offered the chance to buy ticket 2. Would you expect there to be a difference between the rates at which people choose to take up these options? Again, it seems hard to imagine that there might be. The same purchase price gives the same chance of winning the same prize so a reasonable person would expect that any difference between the uptake of the two offers would just be down to chance.

Unfortunately for reasonable people everywhere, psychological research says something completely different. Specifically, that people interpret percentages and natural frequencies differently and that this changes their behaviours and choices—as described below.

6.1.1 Base rate neglect again

Recall, for example, the discussion of base rate neglect in chapter 4. Here it was noted that the way in which probabilities are presented to people has a significant effect on how many of them can solve the base rate neglect problem correctly. Specifically, presenting the probabilities in a table like table 4.2 assists people in determining which need to be compared and how (e.g. Sloman et al 2003). As noted in chapter 4, however, some researchers (e.g. Cosmides and Tooby 1996) go further, arguing that natural frequencies (e.g. 1 in 100) are more easily understood by people and thus produce better results than percentages.

It has been argued that this might stem from natural frequencies being the way in which our minds actually capture information—with percentages being a recent mathematical add on to that process (Gigerenzer and Hoffrage 1998). This implies that percentages are a further level of abstraction away from the way in which we naturally deal with probabilities. Natural frequencies can also, however, retain more information than percentages, which can be important in solving a base rate neglect style problem. For instance, 10 in 1000 and 1 in 100 both collapse to the same probability—1%—but the size of the sample from which observations is drawn is central to understanding how likely, for example, false positives are compared to true positives in a base rate problem.

6.1.2 Disastrous decisions

A similar effect has been observed in research regarding choices made in the face of potential disasters (Welsh et al 2016). Specifically, scenarios described the participants inheriting a house and then the probability of a natural disaster (bushfire or earthquake) in the area upgraded as the result of new scientific information. Some people, however, had the annual probability of a disaster described in terms of natural frequencies (e.g. 1 in 100) while others had them described as percentages (e.g. 1%). This had a marked impact on the decisions that people made, as shown in figure 6.1.

Looking at figure 6.1, one sees that people had four options to select from. These were to: do nothing; take minor or major steps to prepare for the possibility of the disaster; or to sell the property. The responses favoured by people, however, differed according to the way in which the information was presented to them. Specifically, people given the information as a percentage chance of the disaster occurring in each year were more likely to ignore the change in disaster likelihood and select to do nothing, with 27% of all responses falling in this category. In contrast, when the likelihood of a disaster was described in natural frequency terms, only 13% of people

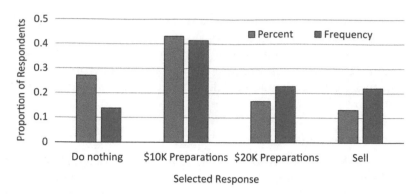

Figure 6.1. Responses to upgraded disaster likelihood by format of information delivery.

selected do nothing as their preferred option. Instead, almost twice as many of this group wanted to sell the property and more were willing to pay for major preparations to the property to mitigate against any impact of the disaster.

These findings suggest that because natural frequencies are more easily understood by people, they are more likely to prompt action. In the experiment described above, of course, the options were carefully balanced such that there was no economic reason to select one over the other but, in real life, there will, in many cases, be an optimal decision and the findings regarding people's reactions to percentages and natural frequencies suggest that people could be 'nudged' towards more appropriate choices via careful selection of the way in which scientific data are presented.

6.1.3 Absolutely twice as bad

Another key concern relating to people's interpretation of probabilities lies in communications regarding risks and whether they are given in absolute or relative terms. For example, the rate of Down syndrome (trisomy 21) in new-born babies increases with the age of their mothers (Newberger 2000), but this can be described in two distinct ways. One is using the absolute rate for a mother of any particular age or the change in absolute risk from a baseline. The more common way, however, is to refer to the relative risk of a mother of one age compared to some baseline (i.e. a younger mother). For instance, a 35-year-old mother is around five times as likely to have a baby with Down syndrome as a 20-year-old and this relative risk increases to around ten times for a 40-year-old mother. In absolute terms, however, the chance remains quite low—increasing from less than 0.1% in a 20-year-old to around 0.5% for a 35-year-old and a little below 1% for a 40-year-old.

Even looking at these numbers while knowing that they portray the same information, most people immediately agree that the five-fold or ten-fold increase seems more concerning than the 0.4% or 0.8% change in absolute risk. Thus, which terminology is used to describe the risk is likely to alter the responses and behaviour of the mother-to-be—including important decisions about what medical tests to undergo in response.

Malenka *et al*'s (1993) research demonstrated exactly this. When two, alternative treatments were described that were equally good in every way but one had its benefits described in relative terms people selected the treatment so described more than twice as often as the alternative that had its benefit described in absolute terms. Specifically, in a scenario where of every 100 people who were infected with a disease 10 were expected to die, a drug that was described as reducing a patient's chance of dying by 80% was preferred to one that reduced the number of deaths expected from a disease from 10 out of 100 to 2 out of 100—an absolute reduction of 8 deaths. The researchers' follow-up questions indicated that the source of the difference was confusion of what the 80% reduction referred to—with most people not adjusting for the expected rate of death and, implicitly, assuming that everyone with the disease was going to die.

This reflects the danger of relative risk assessments—because people can easily misinterpret what exactly the risk is relative to, it leaves the door open to significant errors of belief. It does, however, also suggest that, if you want to convince people to accept a particular course of action based on responding to risk, you are better served by expressing those risks in relative terms. For example, telling people that the annual risk of significant flooding in an area has doubled is far more likely to prompt precautionary action such as purchasing flood insurance than telling them that the risk has changed from 1% to 2%.

6.2 Nudges, defaults and frames

The above are examples of the idea known as 'paternal libertarianism' as promoted by Thaler and Sunstein (2008) in their book *Nudge*. The central idea here is to present information in such a way as to promote decisions that improve the lives of the decision makers, while still ensuring that the person is free to actually make the decision. That is, using what we know about how people respond to information presented in different formats to ensure that we present this information in its best possible light—where best is defined as people making decisions that are in their own self-interest or to the benefit of society as a whole.

6.2.1 Default decisions

Perhaps the most familiar example of this approach is the debate over how best to ask questions about organ donation. Johnson and Goldstein's (2003) paper considered the donation rates across a variety of European countries and found that donation rates, rather than being affected by any presumed cultural differences, were almost entirely driven by the format of the question being asked of people. In those countries where people were asked to opt in to become organ donors, donation rates varied between 4% and 27%, whereas in countries where the default option was to be enrolled and the respondent had to opt out of the program, the lowest enrolment rate was 87.5% and in the majority of countries it exceeded 98%.

Some of this effect could, presumably, be put down to people not reading the form and thus being left with whatever the default option is, but the research in the

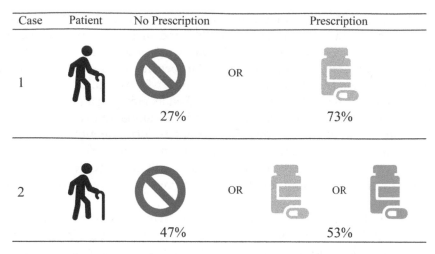

Figure 6.2. Visual representation of Redelmeier and Shafir (1995).

field suggests that it is more than this—simply being the default option does seem to make organ donation more palatable to people.

A similar effect has been shown in doctors' decision making about treatment options. Redelmeier and Shafir (1995) described a hypothetical case where a physician has decided to refer a patient on to a specialist because medications have failed to relieve their chronic hip pain. Having made this decision, they then realise that there was a medication they had not yet tried. A group of physicians was then asked to decide between prescribing the new medication or not prior to the referral. A second group of physicians was given the same question—except that they were told that there were two medications the patient had yet to try. Figure 6.2 shows the experimental design and key result.

Looking at the figure, one sees that, when faced with the singular decision—between sending the patient on without trying a single, new medication, the majority (73%) of physicians indicated that they would prescribe the medication. However, when this was complicated by the inclusion of a second, equally suitable, medication, physicians became far less likely (53%) to decide to prescribe anything. This effect cannot be explained by differences in the medications as the same medication (ibuprofen) was the first option in each case. Instead, it seems that the complexity added by making it into a two-step decision (step 1, should I prescribe; step 2, what should I prescribe?) resulted in physicians sticking with the default position. That is, the decision to which they had already committed—referring the patient on to the specialist without further medication.

6.2.2 Framing decisions

This power of the default has the potential to affect a wide variety of decisions. Specifically, it can affect *which* decisions we make. For instance, if your car is written off in an accident, the default decision is: what sort of replacement car should I buy? This may not, however, be the best decision for you to be making. This is a key aspect of the approach to decision making known as decision analysis (DA; Howard 1968)—in fact, DA's first principle is to ensure that you have correctly 'framed' your decision.

Extending the above car example, for instance, consider what decisions you *could* be making. Rather than 'which car to buy', you could be considering the much broader 'what sort of transportation should I use?' The broader decision is, of course, likely to include far more options and thus has a greater likelihood of you being able to achieve your actual objectives (e.g. save money, save time).

Thus, understanding the frame in which a decision is being considered is important as this will allow you to determine whether you are simply being swayed towards default options when thinking outside the box might turn up superior options.

6.3 Decisions under framing

When 'framing' is spoken of in the heuristics and biases literature, however, it does not, generally, refer to the above problem. Rather, it relates to a specific observation regarding how people make decisions under what is called a 'risky choice' paradigm. That is, where people are faced with choices between options that are certain and uncertain, and which has consequences for both the decisions we make and how we choose to convey information.

6.3.1 Grant application

Imagine that you have put in an application for a research grant. The funding body contacts you to explain that, this year, they have only had two applications (including yours) and, as a result offer you two alternatives:
  a) Accept a $50 000 grant.
  b) Wait for the funding body to decide between the two grants, meaning you will receive either nothing or $100 000.

Which option would you prefer?

Now imagine that you have put in an application for a research grant and received word that you have been awarded the full $100 000 that you applied for. While you are completing the paperwork to accept the grant, however, the funding body contacts you to explain that they have made an error—specifically, they have failed to assess one other, rival grant. They offer you two alternatives:
  a) Accept a $50 000 reduction in your grant amount.
  b) Wait for the funding authority to decide between the two grants, which will result in you losing all $100 000 should they select the rival grant or losing nothing if they select your grant.

Which option would you prefer? Is there a difference between your responses to the two questions?

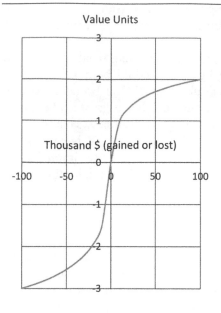

Value Units

Thousand $ (gained or lost)

Key attributes:

1. Diminishing returns for losses and gains.

2. Steeper loss curve.

3. Gains and losses relative to status quo.

Implications:

1. $100 000 is not twice as valuable as $50 000.

2. –$100 000 is not twice as *bad* as –$50 000.

3. –$50 000 is worse than $50 000 is good.

4. Losses ≠ gains.

Figure 6.3. Prospect theory curve.

6.3.2 Prospect theory

Framing was an early area of research in heuristics and biases, with Kahneman and Tversky comparing people's decisions with what economic theories said they should be doing according to expected utility theory. The central conclusions form this work form the basis of prospect theory (Kahneman and Tversky 1979). Figure 6.3 shows a typical prospect theory curve—that is, a curve that describes how a typical person might respond to risky choice problems like the one above—and we will use this as a visual aid to help see the important characteristics of prospect theory. (Note —there are individual differences between people and each person has their own prospect theory value function, which can be determined experimentally through a series of these risky choice problems. Despite this, a majority of people respond as described herein.)

Looking at figure 6.3, one can see that its top-right quadrant bears a strong resemblance to an expected utility curve—in that, as the dollar value rises, so does the value that the person places on it (i.e. it is a monotonically increasing function)— but that this increase is non-linear. Specifically, it shows diminishing returns—where the second $50 000 is less valuable than the first $50 000.

This observation explains a key observation from behavioural studies. Specifically, that people's *risk attitude* changes depending on whether they are dealing with losses or gains. We often speak about a person being 'risk seeking' or 'risk averse' but, according to prospect theory, these labels are more properly ascribed to a person's behaviour under specific conditions—that is, the same person is sometimes risk seeking and sometimes risk averse. One of the key drivers of a person's risk attitude in any given circumstance is whether they are dealing with losses and gains—and this is simply a function of diminishing returns.

Looking at the gains quadrant of figure 6.3, for instance, one can see that a gain of \$50 000 equates to a 'value' of around 1.7. A \$100 000 gain, by comparison, has a value of 2. Given this, a 50% chance of a \$100 000 gain only has a value of 1 (50% of 2), which is less than the value of \$50 000. So, given a choice between a certain \$50 000 and 50/50 chance of nothing or \$100 000, most people select the certain \$50 000. In the losses domain exactly the same effect, however, causes people to become risk *seeking*. That is, because a \$100 000 loss is not twice as *bad* as a \$50 000 loss, a 50% chance of a \$100 000 loss is preferred to a certain \$50 000 loss. This also explains many difficulties that arise around policies that might on the surface seem fair and balanced—for example, taking away a benefit in one area but providing it back through another. While the net result of this may be no change in a person's absolute level of well-being, the loss of their current benefit is weighted more heavily than is the equivalent gain. In general, this causes significant problems for any attempt at redistribution of goods or services because the people who are losing feel that loss more heavily and thus will tend to fight harder to prevent any change—even if that change would be better or fairer overall. All of these are effects of diminishing returns and would, therefore, be predicted under expected utility theory.

Where the prospect theory curve differs from expected utility, however, is that expected utility, being based on rational economic theory, assumes that the dollars being gained have the same value as dollars being lost—and thus that a single curve can describe how a person will convert any dollar amount into a value. Kahneman and Tversky's research, however, demonstrated that this is not the case—people weigh losses and gains differently. Additionally, their work showed that people's internal calculus was based not on the overall dollar value of outcomes but rather on *changes* in outcomes relative to the status quo.

Looking back at figure 6.3, one can see the first of these insights reflected in the different curves for the losses and gains. While both show diminishing returns, the losses curve is steeper, reflecting the fact that most people, when making decisions, weigh losses more heavily than equivalent gains. That is, if you offered someone (with a prospect curve such as this) a coin toss where they would lose \$50 000 if the coin comes up tails and win \$60 000 if the coin comes up heads, they would refuse—despite both expected value and expected utility suggesting that they should take this bet. This is simply because they weight the \$50 000 loss as significantly more important than the \$60 000 gain.

The second point is subtler and the pair of questions given in section 6.3.1 are designed to highlight this. In both cases, the options available to you are the same. You have a choice between a certain \$50 000 grant or accepting a 50/50 chance of getting either nothing or \$100 000. Where they differ, of course, is that the first question describes the monies as a gain from your current state of having no grant money whereas the second describes the two outcomes in terms of losses from an expected state of having \$100 000. Logically, this should not affect the decision that you make—given that the objective outcomes are the same—but the fact that people assess losses and gains relative to the status quo means that most people will interpret the options in the first question as gains and those in the second as losses. As a result, in line with the observation about diminishing returns made above, more people will accept the $50 000 when the situation is described—or *framed*—in terms of gains and more will take the risky option when it is described in terms of losses.

This effect is robust and has been replicated with non-monetary losses and gains. For example, Pieters' (2004) oil spill framing problem, which describes alternatives for dealing with an oil spill in terms of either positive (containment) or negative terms (spread), resulted in the proportion of Society of Petroleum Engineers members who selected the certain option changing from 80% to 14%. That is, the majority (80%) were risk averse when the options were described in terms of containing the oil spill but, when described in terms of the oil spill spreading, the majority (86%) became risk seeking. Similarly, McNeil *et al* (1982) demonstrated that choices between medical treatments were affected according to whether the outcomes were described in terms of the percentage chance of surviving the procedure rather than the percentage chance of dying during the procedure—despite the chances being the same in the two conditions.

The implication of this for communicating risks is clear—how you choose to frame choices has a marked impact on how people are likely to respond. In general, if you describe events so that they sound like losses relative to the current state of affairs, people are more likely to adopt riskier choices. This is true even if it is not a genuine loss. For instance, returning to the grant examples in 6.3.1, even if you had not been *told* that you had won the grant but rather simply *believed* that you were going to win the $100 000, then to be offered $50 000 will still feel like a loss and is likely to result in a more risk seeking attitude.

6.4 First, last and best

The above effects all revolve around surface-level changes to the description of options and how this affects which option people refer. By comparison, in the following section, we consider the effect that changes in the number of options or the order in which options are presented can have on people's choices.

6.4.1 Who to choose?

You are looking for a person to join your trivia team and a friend of yours has suggested two of their friends—John and Jane—as possible options. You ask your friend to describe the two and she responds that John is:

  intelligent, industrious, impulsive, critical, stubborn and envious.

She then describes Jane as:

  envious, stubborn, critical, impulsive, industrious and intelligent.

Who would you rather have on your trivia team?

6.4.2 Order effects

When someone is presented with a list of options or a number of pieces of information, the assumption that rational decision theories make is that the order of that list is irrelevant. If you have the same set of options, then which you prefer should not be affected by whether it is first, middle or last in a list. Likewise, such theories assume that when you receive information, you absorb all pieces equally well.

Studies in human memory, however, have demonstrated that this a flawed assumption—for a number of reasons. First, we know that human memory is limited in its capacity; not in terms of long term storage of information but in terms of how many items can be held in short term memory simultaneously and rehearsed to move them to long term storage. While there are individual differences and domain differences in exactly what this number might be, the consensus is around seven (Miller 1956).

As a result, if someone is receiving information, it is safe to assume that once you have gone past seven options or facts, they no longer have all of the information concurrently available to them. Knowing this, one might expect that short term memory would work like a camera trained on cars driving through a tunnel—the cars being the individual facts or options and the tunnel representing short term memory. The tunnel's length means that only seven cars can be inside it at one time and the camera can only see the ones that are. In order for a new car to enter, one has to exit at the other end—as shown in figure 6.4.

In fact, however, this model is incomplete. While it is true that more recently encountered objects, facts, and so on are better recalled than older ones (the 'recency effect'), as one might expect from a model like this where a new item tends to 'push' older items out of mind—research on memory and decision making has also demonstrated 'primacy' effects (see, e.g. Asch 1946, Ebbinghaus 1902). This refers to either people's superior recall of, or the accompanying greater impact of, items presented first in a list. For example, in the questions in section 6.4.1, the two candidates are described using the same seven descriptors taken from Asch (1946). The order in which they are presented, however, affects people's overall impression. Specifically, whether the first descriptor is positive or negative colours a person's overall impression. That is, John, being described as 'intelligent' first, is regarded in

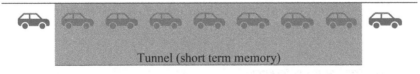

Tunnel (short term memory)

Only cars inside the tunnel can be seen by the camera (short term memory).
For another car to enter, anothermust exit.
More recently added options are more likely to still be in memory.
This explains 'recency' but not 'primacy' effects.

Figure 6.4. (Incomplete) short term memory tunnel analogy.

a better light than Jane, who is first described as 'envious'. This harkens back to the discussion of conversational norms and Gricean maxims in chapter 3 (Grice 1989). Specifically, we expect people to give us the most relevant information they have first —so the order in which the information is given to us *is*, we presume, important.

In recall tasks, however, a similar effect is observed, with people recalling the first item from a list far more easily than later items (generally excepting the last item). A possible reason for this is that people have more time to rehearse an item or information that is given first and, in particular, more time when it is the *only* item being considered—which results in a commensurate increase in the likelihood that it will end up being transferred to long term memory (Marshall and Werder 1972). That is, our memory processes seem to work in a way that reinforces the conversational expectation that whatever information is presented first is going to be the most important.

This observation has probably not surprised many readers—for exactly the reason that we all understand norms of conversation. Even scientific papers, which form a somewhat unusual medium of conversation, generally adhere to our intuitive understanding of such structures. Most start with an abstract in order to put the main finding of the paper front-and-centre, before continuing to the meat of the study and then closing with a repetition of the main finding in the conclusion. That is, papers are written in a manner designed to ensure that readers come away with the main point firmly in mind.

These order effects can, however, undermine the communication of science in some instances. In particular, when attempting to correct a misconception or debunk a myth. How we choose to do so affects the message that the listener takes away. For example, the familiarity backfire effect described by Lewandowsky *et al* (2012) seems to be caused in part, at least, by primacy. This describes a situation wherein someone is attempting to dispel a myth (e.g. the link between vaccines and autism). In doing so, however, they begin by restating the myth before going on to present the counterargument. When the target of the debiasing later recalls what was told to them, however, the first thing that comes to mind is the myth that the speaker intended to dispel. This can actually result in them being more convinced of the myth's truth. Instead, Lewandowsky and colleagues' suggestion is to state the truth first—for example, that there is no link between vaccines and autism—and only then explain the origin of the myth and the facts surrounding it. In this way, primacy works towards the desired outcome of people remembering the truth rather than the myth. This and other methods for countering biases in understanding will be discussed further in chapter 13.

6.4.3 *Comparisons and preferences*

6.4.3.1 *Which would you prefer?*

Imagine that you have won a competition and the prize on offer is a choice between two holiday packages.

1. An all-expenses paid, one-week safari in Tanzania, Africa. Leaving from your home city with all transfers included.

2. An all-expenses paid, one-week tour to the ruins of Machu Picchu in Peru. Leaving from your home city with all transfers included.

Which would you prefer?

Now, imagine that you have won a competition and the prize on offer is a choice between *three* holiday packages.
1. An all-expenses paid, one-week safari in Tanzania, Africa. Leaving from your home city with all transfers included.
2. An all-expenses paid, one-week tour to the ruins of Machu Picchu in Peru. Leaving from your home city with all transfers included.
3. An all-expenses paid, one-week tour to the ruins of Machu Picchu in Peru. All transfers are included but it leaves from a neighbouring city to which you will have to make your own way.

Which would you prefer?

6.4.3.2 Constructing preferences

Looking at the above questions, it is reasonable to expect that, in the first case, people will have different preferences—that is, after weighing up the pros and cons of each, some people will prefer the Tanzanian safari and some the Peruvian tour. Our expectation, however, is that such preferences are stable. So, when faced with the second question, we would expect the same people to make the same choice. This is because the third option that has been added in the second scenario is clearly inferior to the existing Peruvian tour—having the same features but requiring you to travel to another city to begin rather than leaving from your home city—and so no one would choose this.

This logic, however, is based on a misconception regarding people's preferences. As demonstrated by Ariely (2008), using questions similar to those above, people's preferences are easily swayed by the addition of seemingly irrelevant options. This is because, when faced with a decision like the choice between the safari and the tour of Machu Picchu, we tend not to have a clear, pre-existing preference. Instead, we have to *construct* a preference; and doing so is difficult. The trips offer different but significant, possibly once-in-a-lifetime, opportunities and making a decision regarding which of them is better is not a simple task.

What *is* a simple task, however, is to determine that the Peruvian tour leaving from your city is better than the same tour but leaving from another city. As a result, people faced with this decision are more likely to select Machu Picchu than they would be in a straight selection between the two options in question one—simply because they can, with certainty, say that it is better than the option that no one would have picked. This may seem illogical but it is a robust effect, with Ariely having demonstrated it across several domains. In essence, it seems to boil down to a set of relative judgements. We have three options: Tanzania, Peru and sub-par Peru. The pair-wise comparisons between these are shown in table 6.1.

Table 6.1. Pairwise comparisons of preferences.

	Tanzania	Peru	Sub-par Peru
Tanzania	—		
Peru	?	—	
Sub-par Peru	?	Peru	—

Looking at the table, one sees that the pair-wise comparison between Tanzania and either Peruvian option are uncertain—because deciding between them is complex enough that the addition of an extra, unfunded leg of the journey is not sufficient to make one definitively better than the other. For the comparison between Peru and the sub-par Peru tour, however, there is, immediately, a clear winner. As a result, if you were to score the alternatives using these pair-wise comparisons, with 1 for a win, 0 for a loss and 0.5 for an uncertain result, Peru scores 1.5, Tanzania 1 and sub-par Peru 0.5, resulting in Peru being chosen.

This means that a person's preferences between options are dependent not just on how inherently attractive those options are to them, but also on the limitations of their cognitive processing and the specific set of options that they are provided to decide between. That is, it provides even more evidence that the decisions people make are significantly influenced by the way in which options and information are communicated to them.

6.5 Conclusions

Examples like those above serve to highlight some of the peculiarities of people's decision-making processes that do not accord with rational theories. For example, while it may seem reasonable to assume that people understand percentages and frequencies to mean the same thing, the evidence suggests otherwise. Specifically, that people more easily understand natural frequencies.

Additionally, it is clear that people do not value options in the way that expected utility theory suggests they should—with their behaviour instead conforming to prospect theory value curves such as that shown in figure 6.3. This results in them not only weighing losses more heavily than gains but also becoming susceptible to framing effects—changing their behaviour from risk averse to risk seeking depending on whether options are described to them so as to sound positive or negative.

Rational theories of decision making presuppose—as, again might many reasonable people—that our preferences are set and that, as a result, adding new options or changing the order in which they are presented will not change the preferences between already available options or alter the way in which people incorporate the information being communicated to them. As seen above however, the fact that preferences are often constructed 'on-the-fly' rather than existing prior to a person being posed a question means that the way in which people go about the process of constructing their preferences becomes important—and this depends on the limited cognitive processes we can bring to bear. As a result, we become susceptible to a wide variety of biases resulting from limitations in our cognitive processes—including memory, which is the central theme of discussion in chapter 8. Overall, the findings in this chapter argue strongly for our need to understand human decision-making biases. In addition to assisting scientists to recognise or avoid bias in their own decisions, anyone engaged in the communication of science needs to understand how to frame and present results so as to avoid these biases detracting from or distorting the intended message.

References

Ariely D 2008 *Predictably Irrational* (New York: Harper Collins), 20

Cosmides L and Tooby J 1996 Are humans good intuitive statisticians after all? Rethinking some conclusions from the literature on judgment under uncertainty *Cognition* **58** 1–73

Ebbinghaus H 1902 *Grundzüge der Psychologie* vol 1 (Leipzig: Veit)

Howard R A 1968 The foundations of decision analysis *IEEE Trans. Syst. Sci. Cybernet.* **4** 211–9

Johnson E J and Goldstein D 2003 Do defaults save lives? *Science* **302** 1388–9

Grice H P 1989 *Studies in the Way of Words* (Cambridge, MA: Harvard University Press)

Hoffrage U and Gigerenzer G 1998 Using natural frequencies to improve diagnostic inferences *Acad. Med.* **73** 538–40

Kahneman D and Tversky A 1979 Prospect theory: an analysis of decision under risk *Econometrica* **47** 263–91

Lewandowsky S, Ecker U K, Seifert C M, Schwarz N and Cook J 2012 Misinformation and its correction: continued influence and successful debiasing *Psychol. Sci. Publ. Interest* **13** 106–31

Malenka D J, Baron J A, Johansen S, Wahrenberger J W and Ross J M 1993 The framing effect of relative and absolute risk *J. General Inter. Med.* **8** 543–8

Marshall P H and Werder P R 1972 The effects of the elimination of rehearsal on primacy and recency *J. Verb. Learn. Verb. Behav.* **11** 649–53

McNeil B J, Pauker S G, Sox II C Jr and Tversky A 1982 On the elicitation of preferences for alternative therapies *New Engl. J. Med.* **306** 1259–62

Miller G A 1956 The magical number seven, plus or minus two: some limits on our capacity for processing information *Psychol. Rev.* **63** 81

Newberger D 2000 Down syndrome: prenatal risk assessment and diagnosis *Am. Fam. Physic.* **62** 825–32

Pieters D A 2004 *The Influence of Framing on Oil and Gas Decision Making* (Marietta, GA: Lionheart)

Redelmeier D A and Shafir E 1995 Medical decision making in situations that offer multiple alternatives *Jama* **273** 302–5

Sloman S A, Over D, Slovak L and Stibel J M 2003 Frequency illusions and other fallacies *Organ. Behav. Hum. Decis. Process.* **91** 296–309

Thaler R H and Sunstein C R 2008 *Nudge: Improving Decisions about Health, Wealth, and Happiness* (New Haven, CT: Yale University Press)

Welsh M, Steacy S, Begg S and Navarro D 2016 A tale of two disasters: biases in risk communication *38th Annual Meeting of the Cognitive Science Society (Pennsylvania, PA)* eds A Papafragou *et al* (Austin, TX: Cognitive Science Society) pp 544–9

IOP Publishing

Bias in Science and Communication
A field guide
Matthew Welsh

Chapter 7

I'm confident, you're biased: accuracy and calibration of predictions

This chapter focuses on the concept of calibration: a measure of how well our predictions or estimates match up against reality. This is related to how we think about our own thoughts and the extent to which we are accurately able to understand the limits of our own knowledge—that is, our metacognition and traits such as confidence that depend on such tendencies.

In particular, the near-ubiquitous bias known as overconfidence is described as the tendency for us to believe that we know more than we do, particularly under circumstances of high uncertainty. Multiple variants of this bias are described and possible causes discussed along with some related effects.

As noted in chapter 3, when we need to reduce our own uncertainty, we often elicit expert opinion. Continuing this theme, perhaps the main point of this chapter is to highlight a number of reasons that such elicited values need to be taken with a grain of salt—unless the elicitation has been carefully conducted by someone with an eye towards reducing the impact of overconfidence and other biases.

7.1 Confidence: good or bad?

Confidence is a trait that many people aspire to and which is commonly linked to positive life outcomes—increased chances of doing well at study, getting a job or ahead in one's career and so on. When thought about in terms of decision making and biases, however, it should be clear that the most important aspect of confidence should be how well it aligns with a person's actual ability. That is, while we value confidence, we do so because we assume that a person's confidence comes about due to their competence. This, of course, has the natural consequence that a person's level of confidence should vary across different domains wherein they have differing levels of ability.

Self-report measures of confidence, such as the personality evaluation inventory (Schrauger and Schohn 1995), can take this into account, asking a person to rate their own confidence within various fields of activity—academic, sporting, social, etc. There are, however, positive correlations between all such measures, indicating that people who are more confident in one area tend also to be more confident in others. The question is whether this reflects a general correlation between competencies in these different areas or an underlying trait, confidence, that is not dependent on competency per se.

Work on so-called 'online' confidence (measured during another task by asking people after each item to judge how likely they think it is that they answered that item correctly) suggests that there is, in fact, a stable trait of this sort (see, e.g. Stankov *et al* 2012), which reflects a person's overall confidence in their ability to correctly solve problems. In both interpretations of confidence, therefore, it seems likely that confidence will not map precisely onto competency. Rather, they suggest that a person who is confident because of their competency in one area will carry over some of that confidence into areas where they are less competent.

Another concern arises from what Schrauger and Schohn (1995) term 'selective appraisal'. This refers to any effect that would cause a person to misremember or misevaluate their own performance. In addition to flaws in memory, such as those discussed in various places throughout this book and, in particular, in chapter 8, this could include other personality traits such as neuroticism, which might cause one person to focus on their own failures while a second, equally competent, person focuses more on their successes and thus ends up more confident despite equivalent performance.

7.2 Under and over

The above suggests that people's confidence could differ from their competence in two main ways. First, they could be more confident than their competence justifies: overconfident. Or they could be less confident than their competence justifies: underconfident. Importantly, which of these a person might display or the extent to which they display them may still vary by domain. For example, a person with high competence across a wide range of topics may have a justifiably high overall level of confidence. When asked about an area they are less competent in, however, they could either underestimate how much of their knowledge is transferable and thus be underconfident or overestimate and end up overconfident.

The key point thus revolves around whether a person can accurately recognise whether the confidence they feel is justified or not. Research on this point is largely unequivocal: the answer (for most people at least) is *not*.

7.2.1 Ignorance is bliss

An important piece of evidence supporting this somewhat pessimistic conclusion is the Dunning–Kruger effect (Kruger and Dunning 1999). This is the observation that ignorant people are, largely, unaware of the extent of their own ignorance—leading them to believe themselves to be more competent than they are. For example, in the original paper on the effect, participants who scored on the 12th percentile on various tasks believed themselves to have performed at the 62nd percentile—a significant mismatch between their competence and confidence. At the other end of the performance scale, by comparison, a reversal is seen, with more competent people *underestimating* their performance relative to other people. Here it seems that, because the best scorers found the tasks easy, they assume the tasks would have been easy for everyone else as well.

7.2.2 The hard–easy effect

The Dunning–Kruger effect is mirrored in the heuristics and biases literature by the hard–easy effect (Lichtenstein *et al* 1982)—the observation that measures of people's calibration show overconfidence when the questions being asked are hard and underconfidence when they are easy. That is, people estimate that they will be right more often than they actually are when questions are hard, and less often when questions are easy.

This will, doubtless, feel familiar to many scientists faced with communicating with people outside their fields: people who clearly know very little but who are willing to confidently argue their ill-formed positions are almost stereotypical. In contrast, a diffident presentation of a well-founded opinion (with appropriate caveats) often displays far less confidence than that to which an expert's competence might entitle them.

That said, even amongst scientists, overconfidence often remains the more pressing concern, for one key reason: elicitation. As noted in chapter 3, a major aspect of scientific communication is the elicitation of uncertainty—using expert opinion (including that of scientists) to reduce our uncertainty regarding aspects of the world. The places where such expert advice is most needed, however, are generally those where uncertainty is highest and there are no easy answers. That is, elicitors are typically asking difficult questions and so, in accordance with the hard–easy effect, we would expect experts to be overconfident when answering these.

7.2.3 How good a decision maker are you?

How do you think you rank against your peers (i.e. people at your place of work) on each of the following traits?

		Percentile								
		$\leqslant$10th	20th	30th	40th	50th	60th	70th	80th	$\geqslant$90th
1	Driving ability									
2	Generosity									
3	Thoughtfulness									

For each of the following five questions, give a low value and a high value such that you are 80% confident (i.e. would accept 1:4 odds) that the true answer will lie within your range:

1. As of January 2017, how many officially (International Astronomical Union) named moons did Saturn have?

              ( )   to   ( )

2. What is the melting point of vanadium in degrees Celsius?

              ( )   to   ( )

3. How many million years ago did the Jurassic period begin?

              ( )   to   ( )

4. What proportion of the Earth's surface is covered by the Atlantic Ocean (using the International Hydrographical Organization's definition of its extent)?

              ( )   to   ( )

5. In what year was penicillin first used to successfully treat a bacterial infection in humans?

              ( )   to   ( )[1]

Looking back at the five questions immediately above, how many of your ranges do you think will contain the true value?

              ( )

7.2.4 Forms of overconfidence

The above questions reflect three somewhat different forms of overconfidence. The first reflects an effect distinguished as 'overplacement' (Moore and Healy 2008) and which maps closely onto the Dunning–Kruger effect described above. Most people, when faced with the task of estimating where within a particular population they would rank on some (positive) trait will overestimate. Some few, whose performance is actually high may underestimate but, in general, across a group of people, the overall effect is one of overconfidence with a common observation being that more than 80% of people rate themselves as falling above the median.

The second set of questions are designed to demonstrate the form of overconfidence known as overprecision (Moore and Healy 2008), which relates to how often observed values fall within predicted ranges. Given a set of questions like those above, which ask for ranges that you are 80% confident that the true values will fall within, the expectation is that—of you are setting those ranges accurately—80% of those ranges should contain the true value. That is, we would expect well-calibrated people to average four out of five correct as shown in figure 7.1. (*Note*—we do not expect exactly four out of five as the 80% chance applies to each question individually. That is, the appropriate comparison is with a binomial distribution reflecting this 80% chance—with a mean of four but with three or five being fairly likely.)

Across decades of research into this effect (see, e.g. Lichtenstein *et al* 1982, Morgan and Henrion 1990), the evidence overwhelmingly shows, however, that

---

[1] Q1) 53, Q2) 1910 °C, Q3) 201 MY, Q4) 16.6%, and Q5) 1930.

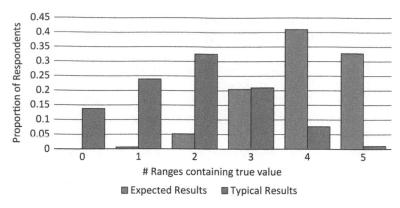

Figure 7.1. Expected distribution for well calibrated people setting 80% confidence ranges across five questions versus typical experimental results.

people are not well calibrated. Rather, the ranges that are elicited from them contain the true value far less often than they predict. That is, they are overconfident. A common observation is that ranges predicted to contain actual values 80% or 90% of the time actually contain them less than half (50%) of the time. Put another way, while people expect values to fall outside such ranges occasionally—the 10% or 20% of results lying outside the predicted range which are referred to as *surprises*—they end up being *surprised* on an additional ~30% of occasions.

Given this, the expectation from the set of five questions in the above section is that, rather than you getting four out of five right—as would be expected if you are well-calibrated—if you show this effect to the same extent as most people, it is more likely that you will get two out of five. (Of course, knowing that you are reading a book about biases means that you are not truly naïve as regards the bias and thus your score could be somewhat *better* than would be expected from someone not thinking about biases when they tried to answer the questions.)

The final question in the above section reflects the third type of overconfidence from Moore and Healy (2008)—'overestimation'. This refers to a general tendency that people have to overestimate the likelihood of getting particular questions correct. If you observe people taking a test of any sort and ask them how likely they are to get each question right or to estimate their overall score, they overestimate compared to their actual performance. There is also evidence that, for example, people overestimate the speed at which they can complete tasks (Buehler, Griffin and Ross 1994)—an idea that will picked up again in the discussion of the 'planning fallacy' later in this chapter.

7.3 Implications of overconfidence

These different forms of overconfidence affect decisions in different ways. Two examples of this are given below—the problem of underestimating uncertainty due to overprecision and the planning fallacy that seems to result from a combination of overestimation of one's ability to complete tasks and overplacement, which prevents people recognising how they will perform relative to other people.

7.3.1 Underestimating uncertainty

The overprecision form of overconfidence is regarded as the most robust (Moore and Healy 2008) and is perhaps the most important for scientific applications as it reflects an underestimation of the uncertainty in a parameter. This is important as, whenever a model is being constructed, uncertainty in inputs needs to be elicited in order to set the ranges of the model's parameters. If the input parameters underestimate the person's true uncertainty (i.e. the ranges are narrower than the expert's knowledge actually justifies) then the model's outputs are likely to mirror this—underestimating the range of possible outcomes.

Sometimes people question the importance of this—arguing that, if the parameter ranges are accurately located, then the model outputs will tend to also produce accurately located best estimates and that the calibration of the range is a secondary concern. This is true, however, only in situations where decisions are being made based solely on the best estimate—without any thought for flexibility. To take a simple example, imagine that you are constructing a geological model to calculate the amount of hydrocarbons in a particular oil field. Given what you know and the model you are using, then if you accurately capture your uncertainty, the range of possible outcomes runs from 100 million barrels (MB) to 700 MB with a best estimate of 400 MB.

If overconfident inputs are used, however, while the best estimate may stay the same (400 MB), the possible range is reduced to 300–500 MB. While this would not change the decision to go ahead with development—as expected reserves of 400 MB are well worth developing— it might change the *way* in which you would choose to develop. Looking at figure 7.2, for example, you can see that the optimal size of the development (number of wells to be drilled, size of platform built, etc.) depends on the expected reserves. With the overconfident assessment of reserves, however, the probability of needing a small or large development is zero.

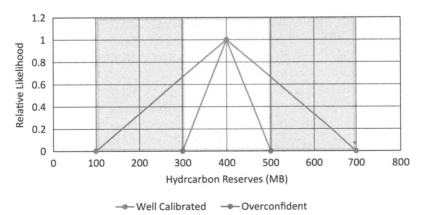

Figure 7.2. Illustrative example of overconfidence in range of hydrocarbon reserves. The coloured areas indicate the appropriate development size: small (red); medium (yellow); and large (green).

If, on the other hand, we had correctly characterised the uncertainty in the reserves, we would know that, while a medium development is the one we most likely need, there is around 22% chance of either a small one or a large one being the best fit for the actual reserves. That is, a 44% chance that the medium-sized development will not be optimal. Given this, the best plan may be something like a phased development, where a small but expandable development is put in place and the outcomes from the initial wells are used to reduce our uncertainty and determine whether additional wells are needed or not. Someone using the over-confident modelling, in contrast, would never come to this conclusion as the medium development is the optimal one for all of the outcomes that they are considering. That is, despite the two models having the same best estimate, how well we understand the uncertainty has significant implications for the decisions that we should make (for a fuller example, see Welsh *et al* 2007).

7.3.2 Plans and planning

Another area in which people's overconfidence has significant real-world outcomes is in what is known as the planning fallacy—the tendency of people to underestimate how long tasks will take them to complete, how much they will cost, and so on. This contributes to the all-too-common experience of deadlines that seem reasonable at first but prove increasingly difficult to meet and the tendency of large projects to suffer time slippage and cost overruns. For example, the Large Hadron Collider was officially approved in 1994 with an expected completion date in 2005 (CERN 2017). However, cost overruns of ~20% pushed this date back to 2007 before additional difficulties resulted in the first experiments not taking place until late 2009—a four year time slippage. This is, however, quite typical of mega-projects, with Merrow (2003) indicating that oil industry mega-projects averaged 46% cost overruns and 28% time slippage.

There are multiple factors that contribute to the planning fallacy—including motivational biases in tendering processes, which encourage underestimation of costs and times—and memory effects relating to the availability heuristic, which will be discussed in chapter 8. Overconfidence can also play a significant role, however, and this seems related to both the overestimation and overplacement effects described above.

As noted above, overestimation refers to people's tendency to believe too strongly in their own abilities—which includes the tendency to underestimate how long a task will take them to complete and how easily they will be able to do so. The corollary of this, of course, is that tasks (particularly complex ones) are more difficult than people believe they will be and tend to take longer to complete—or, sometimes, cannot be completed, meaning that goals need to be reconsidered. This results in increased costs due both to additional time costs but also unforeseen work that needs to be done or information that must be gathered after the project has begun.

The role of overplacement on the planning fallacy is, perhaps, subtler, connecting to the idea of benchmarking. That is, given that people have often seen other people undertaking tasks and are well aware that *other* people fall victim to the planning fallacy (Buehler *et al* 1994), the best explanation for why they believe they will not is that they tend to believe that they have superior abilities relative to those people. This tendency of people to think that they are superior to most other people (overplacement) results in them being able to justify to themselves their over-estimation. That is, the fact that the vast majority of projects take longer than the people involved said they would is not relevant to *my* project because I am smarter, more diligent and better prepared than the vast majority of people. In fact, of course, such confidence is rarely warranted and thus people's tendency to be overconfident in these three distinct yet related ways results in people vastly underestimating the uncertainty in their own estimates of the time, cost and likelihood of success of projects—including scientific or engineering projects.

7.4 Reducing overconfidence

Given the near-ubiquity of overconfidence and the range of tasks which it affects, it will come as no surprise that reducing bias from overconfidence has become a major area of interest in judgment and decision-making research. While it has been demonstrated that making people aware of their overconfidence can reduce its level (see, e.g. Lichtenstein *et al* 1982, Welsh *et al* 2006) such work also demonstrates that the correction that this causes is insufficient to compensate entirely for the effect. This is because such an intervention does not address the underlying causes of over-confidence. (*Note*—the original explanation proposed by Tversky and Kahneman (1974) for overconfidence in range estimates is that people anchor on their best estimate and thus produce end-points of their ranges that are too close to the best estimate and thus ranges that are too narrow, leading to overconfidence. While often repeated, this theory has not held up to rigorous testing, suggesting that over-confidence is primarily caused by other effects; see, e.g. Block and Harper (1991).)

For instance, telling people that their estimated ranges are too narrow will cause them to provide wider ranges, but they often do not adequately understand by how much these ranges need to be widened to account for their true level of uncertainty. When this is explained to them, in fact, they often baulk at providing ranges that would be wide enough to accurately reflect their uncertainty—instead generating ranges that are informative as to what their best guess is. This effect was described by Yaniv and Foster (1995) as the informativeness–accuracy trade-off (IAT) wherein people's preference is for both generating and receiving ranges that are informative about the estimator's best guess rather than accurately capturing their uncertainty. This is an important consideration for anyone interested in elicitation of unbiased information: peoples' preferred cognitive processes are strongly held and not easily overcome. As a result, it is important to consider whether there are alternative ways of asking for the same information that can avoid these tendencies.

7.4.1 Format changes

Taking this tack and recalling the sometimes strange effects of changing the way in which we ask questions of people described in chapter 6, there are a number of methods by which overconfidence can be reduced. For example, when considering overprecision in range estimates, a demonstrated benefit can be observed by having people evaluate rather than generate a range (Winman *et al* 2004). That is, rather than asking them to produce a range that they believe will contain the true value to a stated degree of confidence, instead, provide them with a range and ask them how likely it is that the true value will fall within that range. People find this second task more natural and their responses are, as a result, better calibrated.

Another demonstrated method for debiasing range estimates is to more strictly control how those ranges are generated. Specifically, rather than asking for an 80% range, for example, one can ask a person to give a value that they are 90% certain the true value will fall above and, once that estimate is received go on to ask for a value that they are 90% certain that the true value will fall below (Juslin *et al* 2007). These two values—the 10th and 90th percentile of the person's subjective probability distribution—are, logically, equivalent to their subjective 80% confidence interval. However, the fact that the person is conducting two distinct tasks—one searching for feasible low values and one for feasible high values—results in better estimates of both end-points, seemingly as a result of not having to simultaneously consider high and low values, thereby enabling the person's full cognitive resources to be applied to each question rather than them attempting to create the two estimates constituting a range simultaneously. It also neatly side-steps the IAT described above as, rather than using their range to tell you where they think the true value is most likely to fall, they become focused on telling you where each *end-point* is most likely to fall.

7.4.2 Inside outside

Another means of reducing overconfidence—this time related to the planning fallacy —is through a perspective shift referred to as taking the 'outside view' (see, e.g. Lovallo and Kahneman 2003). People's natural tendency is to regard each project or task as unique and consider it in isolation—taking an 'inside' view. This, however, ignores potentially relevant information available from looking at similar projects undertaken by other people—their rates of success, time to complete and costs—and results in overconfidence due to the above-described tendencies of people to regard themselves as more able than they actually are in both absolute and relative terms.

Of course, scientists can, with some justification, make claim to being above average in a number of ways. However, a key thing to remember is that, when considering the reference class for this 'outside' comparison, is that this comparison set is composed of estimates made by similarly gifted people. That is, while you might be justified in believing that your estimates are better than a person on the street, those are not the people against whom your estimates are being measured.

For example, Kahneman and Tversky (1982) describe a personal experience of this effect while taking part in the design of a new curriculum. The team thought that they were on track for successful and timely completion of the task within 30 months (their highest estimate) until they thought to ask a team member who had seen a number of similar tasks undertaken how long they typically took. This team member, on reflection, realised that 40% of teams *never* completed their curriculum design tasks and that, of those who did, none had done so in less than seven years. They also concluded that the current team was not, in fact, particularly special—

being below average in terms of their expertise and resourcing—making it a line-ball call as to whether they would actually finish.

The team ignored this advice and pressed on—finally completing the task eight years later. The curriculum was, then, set aside and never used.

The upshot of this is clear. Even experts who understand the errors that can affect decision making tend not, naturally, to think about tasks in a way that results in the best possible understanding of how a project is likely to proceed. Rather, people have a tendency to focus on the individual characteristics of a project and convince themselves of its uniqueness—a tendency likely to be exacerbated by effects like confirmation bias (chapter 5), which result in our seeking out evidence that supports our overly optimistic estimates.

7.4.3 Elicitation tools and processes

The above insights into how overconfidence affects estimates and methods for reducing this has informed a number of elicitation methodologies—that is, pre-scribed methods for asking experts to make estimates so as to avoid or limit biases that go further than simple advice to make ranges wider or changes to question formats. For example, Haran *et al*'s (2010) SPIES procedure, which divides the total range of possible values into segments and requires that participants rate the likelihood of the true value falling within each of these. This, of course, is more time-consuming than simply asking someone to estimate a range but prevents people ignoring many feasible values to focus on just their best estimate, and the tendency to give estimates that are informative about the location of the best guess rather than the range of possible values.

The computerised MOLE process (see, e.g. Welsh and Begg 2018) has similar underlying ideas—starting with a very wide range from which are drawn randomly chosen pairs of values. These are presented to the elicitee and they select which they believe is closer to the true value and how confident they are in their choice. The program then selects additional pairs of values while reducing the range to be consistent with the person's confidence judgements (e.g. if they are 100% confident that the true value is closer to 100 than 300, this logically entails that values above 200 do not need to be considered further). Finally, the person's choices and confidence ratings from the set of judgements are combined to produce a final probability density function across the values they believe are feasible. This process: requires the elicitee to consider values from across a wider range than they might otherwise consider; allows them to make relative rather than absolute judgement (which people find easier and are better at); enables the elicitee to be asked to make multiple judgements regarding the same parameter while preventing them from simply repeating their best estimate (thus drawing on the wisdom of crowds effects as discussed in chapter 3); and, finally, it reduces the impact of any individual anchoring values (chapter 8) by forcing consideration of multiple values.

By assisting people to make decisions in a way that avoids the natural tendencies that lead to bias, these techniques markedly improve calibration. Specifically, they result in far less overconfidence than is typically observed in range estimation tasks and less even than is seen when elicitees have been made aware of the bias and been asked for their estimates using appropriately formatted questions.

7.5 Conclusions

The key observation of this chapter is not that confidence is bad or that people are always biased (although I could, with some confidence, have removed the comma from this chapter's title). Rather, what I hope you will take way from this is that, while the ways in which we tend to think do result in predictable, systematic biases, we are capable of deliberately thinking in other ways and that doing so can reduce the impact of these biases on the estimates that we make.

That is, the goal is not simply to make you aware of the existence of biases resulting from miscalibration in our levels of confidence—the observation that people are almost always overconfident when dealing with highly uncertain events. While this is important information to have—particularly when eliciting estimates from other people or when providing estimates in response to others' questions—the more important observation is that these effects arise from the cognitive processes to which our minds default and we need to recognise that these default processes are *not* the only ones available to us. When we do so, this allows us to ask questions of ourselves and others in alternative ways and limit the bias in our estimates.

Being confident is a good thing, so long as that confidence is justified.

References

Block R A and Harper D R 1991 Overconfidence in estimation: testing the anchoring-and-adjustment hypothesis *Organ. Behav. Hum. Decis. Process.* **49** 188–207

Buehler R, Griffin D and Ross M 1994 Exploring the "planning fallacy": why people underestimate their task completion times *J. Person. Social Psychol.* **67** 366

CERN 2017 CERN timelines https://timeline.web.cern.ch/timelines/The-Large-Hadron-Collider (Accessed: 17 October 2017)

Haran U, Moore D A and Morewedge C K 2010 A simple remedy for overprecision in judgment *Judgm. Decis. Making* **5** 467

Juslin P, Winman A and Hansson P 2007 The naive intuitive statistician: a naive sampling model of intuitive confidence intervals *Psychol. Rev.* **114** 678

Kahneman D and Tversky A 1982 Variants of uncertainty *Cognition* **11** 143–57

Kruger J and Dunning D 1999 Unskilled and unaware of it: how difficulties in recognizing one's own incompetence lead to inflated self-assessments *J. Person. Social Psychol.* **77** 1121

Lichtenstein S, Fischhoff B and Phillips L 1982 Calibration of probabilities: the state of the art to 1980 Judgement Under Uncertainty: Heuristics and *Biases* ed D Kahneman, P Slovic and A Tverski (Cambridge: Cambridge University Press) pp 306–34

Lovallo D and Kahneman D 2003 Delusions of success *Harv. Bus. Rev.* **81** 56–63

Merrow E W 2003 Mega-field developments require special tactics, risk management *Offshore* 6 January www.offshore-mag.com/articles/print/volume-63/issue-6/technology/mega-field-developments-require-special-tactics-risk-management.html (Accessed: 18th October 2017)

Moore D A and Healy P J 2008 The trouble with overconfidence *Psychol. Rev.* **115** 502

Morgan M G and Henrion M 1990 *Uncertainty: A Guide to Dealing with Uncertainty in Quantitative Risk and Policy Analysis* (New York: Cambridge University Press)

Shrauger J S and Schohn M 1995 Self-confidence in college students: conceptualization, measurement, and behavioral implications *Assessment* **2** 255–78

Stankov L, Lee J, Luo W and Hogan D J 2012 Confidence: a better predictor of academic achievement than self-efficacy, self-concept and anxiety? *Learn. Indiv. Differences* **22** 747–58

Tversky A and Kahneman D 1974 Judgment under uncertainty: heuristics and biases *Science* **185** 1124–31

Welsh M B, Begg S H and Bratvold R B 2006 Correcting common errors in probabilistic evaluations: efficacy of debiasing *Society of Petroleum Engineers Annual Technical Conference and Exhibition (January)*

Welsh M B and Begg S H 2018 More-Or-Less Elicitation (MOLE): reducing bias in range estimation and forecasting *EURO J. Decis. Process.* (Special issue: Advances in Behavioural Research on Supported Decision Processes) accepted

Welsh M B, Begg S H and Bratvold R B 2007 Modelling the economic impact of common biases on oil and gas decisions *Society of Petroleum Engineers Annual Technical Conference and Exhibition*

Winman A, Hansson P and Juslin P 2004 Subjective probability intervals: how to reduce overconfidence by interval evaluation *J. Exp. Psychol.* **30** 1167

Yaniv I and Foster D P 1995 Graininess of judgment under uncertainty: an accuracy-informativeness trade-off *J. Exp. Psychol.* **124** 424

IOP Publishing

Bias in Science and Communication
A field guide
Matthew Welsh

Chapter 8

Sub-total recall: nature of memory processes, their limitations and resultant biases

This chapter examines in greater detail the role of memory on human decision-making biases and errors in communication. Specifically, how the peculiarities of human memory affect the way in which we record, process or retrieve information. To understand how and why these biases occur, it is necessary to understand—at least at a surface level—the different processes that, together, compose what we think of as 'memory'.

Understanding how these processes tend to operate enables us to understand when and why people forget or misremember things, which is important as these errors form the core of a wide array of biases including: hindsight bias; primacy and recency effects; and biases arising from the availability heuristic. They also have clear, if possibly less central effects, on a variety of other biases including anchoring and overconfidence as discussed below and elsewhere in the book.

8.1 Remember this

Amongst researchers who study memory, there are a number of key distinctions drawn between different types or components of memory. First, and most familiar, is the distinction between short term and long term memory. Short term memory refers to a person's ability to hold a number of different items or examples of some event in their conscious thoughts simultaneously and is limited to approximately seven items for most people and most types of item (Miller 1956). This is sometimes regarded as almost synonymous with working memory (Baddeley and Hitch 1974)—although the later actually includes other, non-mnemonic cognitive processes like fluid intelligence used in the manipulation of 'objects' in a person's short term storage.

Long term memory, by comparison, refers to the set of all memories that a person has stored—including both episodic memories (our autobiographical memories of events and their order) and semantic memories of facts or other, distinct, pieces of information (Tulving 1972). This information exists, for the most part, in our unconscious minds excepting only when information from long term memory is called back into conscious thought.

Thus, the connection between these two types of memory—the processes used to move memories from one type of memory to the other—are also essential to understand. For instance, a key observation is that moving a memory into long term storage depends on rehearsal of the information held in short term memory (see, e.g. Marshall and Werder 1972) and a process known as elaboration, through which the memory is integrated into schema of other facts. However, these processes are affected by the emotional importance of the memory (Hamann 2001). That is, more emotionally charged memories—being better able to hold one's attention and encouraging longer rehearsal times and further elaboration—are more likely to be moved to long term memory.

The final core memory process that needs to be considered is retrieval—the method by which we search our long term storage for information and then recall it to our conscious, short term memory (see, e.g. Atkinson and Shiffrin 1968). This is an undeniably complex process with many components—as anyone who has ever struggled to recall some information only to have it come easily to them at a later time will attest—depending not just on the memory itself and its emotional attributes but also on the context in which it was learnt and the other memories to which it has been linked. For our purposes, however, it is sufficient to understand that this is the case without concerning ourselves too greatly with exactly how this is best explained.

8.1.1 Forget about it!

Of course, no discussion of memory could proceed without some discussion of memory's twin, forgetting. Everyone has had the experience of being given some new piece of information—a PIN, phone number, or a person's name   and then, shortly thereafter being unable to remember it. Similarly, memories that have not been accessed in any way over a long period have a tendency to be forgotten (whether temporarily or permanently). Consider, for example, the following set of questions:

1. What did you have for dinner last night?
2. What did you have for dinner 7 days ago?
3. What did you have for dinner one month ago?
4. What did you have for dinner one year ago?

Unless you keep a food journal (or a very detailed Instagram or Facebook account), the basic expectation regarding the accuracy of your memory is that you are extremely likely to remember the correct answer to the first question, perhaps struggle somewhat to recollect an answer to the second, struggle more to answer the third and, finally, be extremely unlikely to recall what you had for dinner a year ago (unless, of course, today is your birthday or other significant, annual event—in which case: congratulations!).

Memories that are more important than 'dinners' will, of course, increase the likelihood of recollection, but even then, there is good evidence that older memories are more likely to be forgotten. This is reflected in so-called 'forgetting curves' (Anderson and Schooler 1991)— which detail the likelihood of a memory being 'found' based on how important it is, how long ago it occurred and when it was last recalled.

Anderson and Schooler's explanation for this is that given the necessarily limited capabilities of human memory (i.e. acknowledging that perfect recall of all facts and events would require so-much cognitive processing as to make it a huge drain on a person's ability to function in other ways) we have memories that are adapted for the information structure of the environment. That is, they argue that, given limited mnemonic processing, the best way of dealing with a changing environment *is* to forget older and less relevant information.

For example, remembering the names of all of the children in your high school English class may have been important when you were in high school but, now, it is far more important that you remember the names of current colleagues. So, given limited recall, it is better to forget the older information. More important information (like the names of close friends, partners and spouses you may have had) is protected by the fact that it is accessed regularly and is thus less likely to slide off the edge of the forgetting curve.

8.2 Available memories

With this very brief crash course in memory under our belts, we can turn to consideration of how our memories affect the decisions that we make and what, if anything, we can do about it.

8.2.1 Shark versus car

In Australia in 2016, 14 428 people died from the three causes shown in figure 8.1. Please indicate how many of these deaths resulted from each cause.

8.2.2 How many and how often?

While a seemingly strange and arbitrary task, the above is, in fact, the sort of problem that we need to be able to solve to function in a natural environment. Specifically, we need to be able to judge how common or rare different types of events are in order to plan for future outcomes. For example, if you have funding to attend only one of the two conferences you usually attend, you might contemplate which of these is more likely to include talks that you will find interesting.

The process by which we make such judgements is, thus, dependent on our memory. That is, how many events of different types we can recall having seen or heard about. In the case of selecting which conference to attend, you would think back to your previous experiences of the two conferences (or other people's descriptions, etc.) and judge which contained more interesting presentations. That

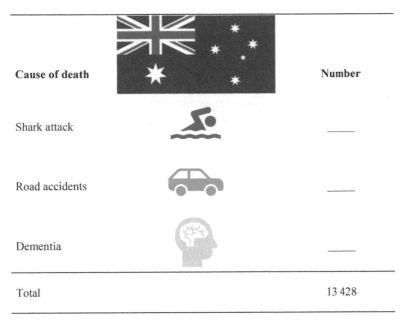

Cause of death		Number
Shark attack		___
Road accidents		___
Dementia		___
Total		13 428

Figure 8.1. Causes of death in Australia, 2016.

is, call to mind the presentations that you recall from each and decide whether they were interesting or not.

This process was described by Tversky and Kahneman (1973) as 'availability'—as observations of people indicated that people judged events to be more likely when more instances of such events could be called to mind. That is, events more easily *available* to memory were judged to be more likely. A process like this makes perfect sense, of course, when operating in a natural environment: if you see an event often, then you should have more instances of it in memory and thus be able to call more of them to mind, leading you to judge common events to, in fact, be common.

As noted in the sections above, however, our memories are not perfect, unbiased recordings of everything we have seen. Some events are more memorable that others for reasons other than their frequency—particularly emotional events, for example, and more recent ones—so the ease with which they are recalled depends not just on the number of instances that you have seen but also on a variety of contextual factors. For example, discussion of a risk raises people's awareness of it, makes instances more available to memory and thus prompts people to overestimate how often events of that type actually occur (see, e.g. Slovic *et al* 1981). (This holds true even for events the person has *never* observed—as people are quite capable of imagining possible events and, to the extent that an event is easily *imaginable*, this has the same result as an event being memorable.)

Another problem for those of us interested in accurately estimating the likelihood of events is that the environment in which we live no longer resembles a natural one in terms of information flow. This means that we need to consider how information is being communicated to us and, more importantly, why. As noted above, the cognitive heuristic called availability makes sense in an environment where you are relying on your own observation of the number of events of different types. Today, however, we live in an information-rich environment wherein different media sources are competing for our attention (typically in order to sell advertising) and the most effective way of doing so is to present events that are rare or unusual and which evoke an emotional response, rather than presenting an accurate reflection of the number of events of different types that are occurring.

The questions in the above section are designed to highlight the possible biases that can come about from this sort of biased information flow. The three possible causes of death are listed in order of increasing likelihood: shark attacks killed 2 people in Australia in 2016 (TCSA 2016), road accidents 1300 (BITRE 2016) and dementia 13 126 (ABS 2016). People seeing these sorts of questions, however, typically overestimate the likelihood of the events that are more commonly covered on the news. That is, they significantly overestimate the likelihood of shark attacks, overestimate the likelihood of road accidents and underestimate the likelihood of deaths resulting from dementia. This is simply because the rarity and emotional impact of shark attacks (plus their ease of imaginability provoked by horror movies) makes them more newsworthy than road accidents, which are, in turn, more impactful for the general populace than one, generally older, person dying as a result of dementia.

That is, media coverage works in exactly the opposite way to a natural environment and, as a result, biases our judgements of likelihoods. This results in people overestimating the chance of death from rare events like terrorism and underestimating far more common dangers like falling off ladders. They then act accordingly—cancelling travel plans to stay home and paint the ceiling.

This multiplies the effects of availability that occur simply as a result of our imperfect memories. That is, our tendency to overestimate the likelihood of events that we have seen recently, have affected us personally or which triggered a marked, emotional response. Simultaneously, we recall fewer very common events than actually occur—because their very ubiquity means that they are not important enough to be rehearsed and thus committed to memory. The result of this is to compress people's probability estimates towards the centre and causing us to react inappropriately to risks.

8.2.3 Availability bias in science

The relevance of this effect to the pursuit and communication of science should be clear. First, it means that, when communicating risks to people, we need to understand that how they will interpret those risks depends not just on how the numbers are presented to them (as described in chapter 6), but also on their own, personal experiences of similar events.

Interestingly, these effects seem to hold true even when the *probability* of an event is described to people—with people reacting more strongly to events that are more personally salient to them. That is, they act as if these salient events are more likely, even when they have exactly the same described probability (see, e.g. Welsh *et al*

2016). For example, when describing the risk associated with a natural disaster (bushfire or earthquake) to people, a key predictor of which mitigation strategy they selected was whether they had previously experienced a disaster of that type, with people facing an unfamiliar disaster type being significantly more likely to choose to do nothing.

Availability can also, however, have an effect in terms of the conduct of science. For example, the observation that people are disproportionately likely to remember surprising events means that we are susceptible to misinterpreting the strength of evidence for and against different positions. This recalls the false consensus effect described in chapter 5, where people deliberately seek out people who agree with them and thus end up believing that there is a consensus agreeing with their own views. The availability effect, in contrast, suggests that people, on seeing an unexpected or surprising result, will encode that event more strongly in memory and, as a result, when they come to assess the evidence for and against a hypothesis, will conclude that the evidence for the more surprising result is stronger than it objectively is.

This tendency is likely to contribute to publication biases (discussed in greater length in chapter 11) where surprising or interesting results are often more easily published (Ferguson and Heene 2012) and, then, more easily recalled by scientists who read them—resulting in an overestimation of the strength for evidence for that surprising result.

This, combined with a predisposition against publishing negative (null) results, has resulted, in fields like psychology, in well-known theories based on a few surprising results that continue to avoid falsification. Ferguson and Heene (2012) describe these as 'undead' theories: 'ideologically popular but with little basis in fact'.

Finally, this bias towards the interesting or unusual can also exacerbate problems with the communication of science—with science journalists acting just as the media typically does and picking up on such results and broadcasting them so they enter the public consciousness as 'a new fact' rather than as the less concise but possibly more accurate 'one, surprising and yet to be corroborated result at odds with decades of solid research'.

8.2.4 Planning fallacy (again)

Another way in which science can be affected by bias from the availability heuristic is the planning fallacy, which the keen reader will recognise from the previous chapter (chapter 7). Recall that this fallacy describes the tendency of people to underestimate how long it will take them to complete tasks and how long those tasks will take—estimation process that are just as relevant to the planning of scientific projects as any others.

In chapter 7, of course, this was discussed in terms of people's overconfidence—leading them to overestimate their abilities relative to other people, for example. There is also, however, strong evidence that availability affects people's estimates of the likelihoods of different events and, as a result, time and cost estimates.

Table 8.1. Fault tree adapted from Fischoff *et al* (1978) showing students' and mechanics' estimates of relative likelihoods for a car not starting given complete and 'packed' fault lists. The 'combined other' value adds the additional faults from the complete list to 'other' to provide a logically equivalent comparison with the 'other' in the packed list.

	Groups			
Possible faults	Students		Mechanics	
Battery	.264	.432	.410	.483
Fuel system	.193	.309	.096	.229
Engine	.076	.116	.051	.073
Starting system	.195		.108	
Ignition system	.144		.248	
Mischief	.051		.025	
Other	.078	.140	.060	.215
Combined other	.468	.140	.441	.215

8.2.4.1 Faulty trees

Table 8.1 provides the classic demonstration of this from Fischhoff *et al* (1978)—now commonly referred to as the unpacking effect. In this experiment, students and mechanics were asked to assess the relative likelihoods of differing reasons for a car not starting. The experimental manipulation was in the number of explicit options they were given—a list of either three or six specific problems and a catch-all 'other' problem category.

Logically, this entails that the 'other' category in the condition with only three explicit categories also includes the three missing explicit categories from the complete list of six used in the other condition. Comparing the 'combined other' values at the bottom of the table, however, one sees that there is a marked difference between the estimates made by participants in the different conditions. Specifically, those who saw the fault list that had only three explicit faults (i.e. where the other three are 'packed' into the 'other' category) made estimates of the combined likelihood of all of these problems that are significantly lower than the combined likelihood assigned by people who were explicitly given these faults in their lists.

In fact, looking across the table, one sees that the general pattern of results seems to be for people to assign weight based on how many categories they have been given —with people assigning more weight to all of the explicitly available categories when confronted with a packed list—more, that is, than would be predicted based on people's estimates of those same categories presented in a fuller list of possible faults.

This finding reflects the common wisdom: out of sight is out of mind. Even the mechanics, who should (one presumes) have a better idea of the true likelihoods of each reason for a car not starting, concentrate their estimates on the explicitly available categories. This results in the total likelihood of the 'combined other' category increasing from 21% to 44% when the 'other' category is 'unpacked' into more, explicitly stated categories.

To put this another way, implicit categories—such as those contained within a general or catch-all category—are less *available* to people's conscious thoughts. The result of this is that it becomes harder to draw examples or other information from memory regarding these and, therefore, this results in less accurate estimates.

8.2.4.2 Planning for problems

This effect can amplify the planning fallacy when, for whatever reason, people start their consideration of tasks at a general rather than specific level. For instance, thinking about how long 'experiment 1' will take to complete rather than considering how long is required for all of the tasks that might compose that experiment—for example, 'design experiment', 'get ethics/safety approvals', 'set up testing equipment', 'recruit participants or prepare materials', 'conduct pilot testing', 'conduct experimental testing', 'analyse data' and 'write up results'.

As a rule, thinking about each of these sub-tasks separately and then summing these is likely to deliver better estimates of the time actually required to complete the overall task than direct estimates of that time—simply because thinking about each will activate (make available to conscious thought) a different subset of a person's knowledge while, at the same time, greatly increasing the total amount of cognitive processing being undertaken, an important consideration given the known limitations of human working memory.

A key point to recall is that expertise does not protect against the contribution of the unpacking effect to the planning fallacy. For example, in a study (Welsh *et al* 2010) that asked participants ranging from undergraduate students to industry engineers to estimate the amount of time lost to various problems in an oil well drilling scenario, engineers made amongst the lowest (and least accurate) estimates when simply asked for 'time lost to all problems'. Only when this general category was broken down into a set of specific problem types did their experience allow them to recognise the potential problems—resulting in their estimates increasing to amongst the highest (and most accurate).

8.3 More or less?

Answer the following questions before reading on.

1. (a) Is the tallest mountain in the UK (Ben Nevis) higher or lower than 2017 m?
   (b) How high would you estimate Ben Nevis to be? _____
2. (a) Is the longest river in Africa (the Nile) longer or shorter than 3425 km?
   (b) How long would you estimate the Nile to be? _____

3   (a) Is the lowest temperature ever recorded on Earth above or below -45 degrees C?

        (b) What would you estimate the lowest recorded terrestrial temperature to be? _____

4   (a) Is the population of North America more or less than 12% of total world population?

        (b) What would you estimate the population of North America to be (% of world)? _____

8.3.1 Anchoring searches

The idea above—that presented information affects what other information a person can recall—is also key to understanding one of the most famously robust decision-making biases, anchoring. Introduced by Tversky and Kahneman (1973) as the 'anchoring-and-adjustment heuristic', it describes the tendency of people's estimates to resemble recently seen numbers—no matter how relevant those numbers are to the estimation task at hand.

For example, Tversky and Kahneman's original demonstration of the effect use a spinning wheel which the participants believed generated random percentages (multiples of 5%) but was, in fact, rigged to always land on either 40% or 25%—the 'anchoring' values. After spinning the wheel, participants were asked to decide whether the proportion of African countries in the United Nations was greater or less than this 'random' number. Finally, they were asked to estimate the proportion of African nations amongst the UN member states. The finding (since replicated across hundreds of studies in a multiplicity of domains) was that people's estimates were affected by the anchoring value that they were shown—despite their belief that this number was random. That is, people anchored on the random number when starting their estimation process.

The four questions above are designed to show this same effect. In each case, the number in the more-or-less part of the question is likely to act as an anchor. In two questions (Q1 and Q4), the anchoring values are ~150% of the true value and are expected to result in estimates that are too high (i.e. above the true value) whereas, in the remaining questions (Q2 and Q3), the anchoring values are 50% of the true value—which is likely to result in estimates being too low (below the true value). The true values for the four questions are: 1345 m, 6853 km, −89.2 C, and 7.8%.

Research on this effect (for an overview see, e.g. Furnham and Boo 2011) has made clear that Tversky and Kahneman's original naming of the underlying heuristic (anchoring and adjustment) is correct in positing two related but distinct processes. The original idea was that people would anchor—that is, start their estimation—at the anchoring value and then consciously adjusting away from there until they reached what seemed to them to be a reasonable estimate. Figure 8.2 shows this process and how it results in bias, with people stopping their adjustment process once they reach what they deem is a feasible response (noting that, if the anchoring value lies within a person's region of uncertainty, they could simply select the anchor as their estimate).

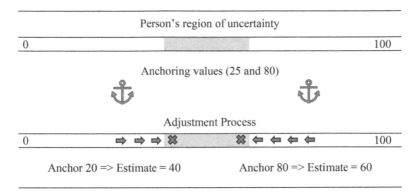

Figure 8.2. Visual depiction of bias resulting from anchoring-and-adjustment. Starting at the anchoring value, a person adjusts until they reach their 'region of uncertainty' (which contains the values they believe are feasible) and then stops. Which anchor they see determines which end of this region their estimate will fall within.

This process, however, does not account for anchoring under all conditions and, in these additional circumstances, memory plays a larger role. This, alternate pathway to anchoring bias is regarded as a form of 'priming'—the tendency of stimuli to 'prime' related concepts or memories, making them more easily available for recollection. The basic priming effect, for instance, as shown in Meyer and Schvaneveldt (1971) asked people to look at pairs of letter-strings and decide whether both strings were actual words or if one or both were made up. For example, the pair BREAD:BUTTER versus the pair BREAD:BODDER. The take away from this research was that people were faster at identifying word pairs when the words were semantically related, for example BREAD:BUTTER, rather than unrelated, such as BREAD:NURSE. That is, the person's memory search for the word BREAD results not just in finding that specific word but also in the activation of related concepts, making their subsequent search for BUTTER faster.

An effect of this kind is also seen in anchoring tasks where participants, on seeing an anchoring value, start by considering whether it is possible that the anchor *is* the correct response and, if not, then start to draw or construct possible values from memory for comparison (Chapman and Johnson 1999). However, the fact that people start with this confirmatory hypothesis testing approach means that the information activated by the initial search is that which has the greatest potential for showing the anchor and the target value to be similar. That is, the anchor primes people to be more readily able to think of reasons that the anchor is a reasonable value, which results in estimates that are similar to the anchor.

8.3.2 Robust and resistant

Anchoring is a core challenge for anyone interested in avoiding or reducing bias in their own behaviour or resulting from their communication of information—simply because it is ubiquitous, affecting novice and expert alike, and has been found to be extremely resistant to attempts at debiasing. For example, anchoring has been shown to affect people's estimates even after the anchoring effect has been explained to them (see, e g., Welsh *et al* 2006, Wilson *et al* 1996). This implies that the cognitive processes underlying anchoring are so fundamental that people cannot help but use them and that conscious control may be impossible.

As a result, any time you are making an estimate, that estimate is likely to be affected by numbers that are at hand—whether reasonable or not—and, as a result, end up too close to the anchoring number. This is of concern even in situations where the anchoring value *is* reasonable—the value of a parameter observed in a previous experiment, for example, that is being used to estimate the value that should be assumed in a new experiment running under different conditions. This is because, as noted above, our natural tendency is to seek reasons that anchoring value could be right and, if it is a previously observed value, then the tendency may be to simply accept it as reasonable without full consideration of the ways in which the current experimental conditions differ from the past.

Anchoring can also play a significant role in communication—as is understood by good negotiators. The first person to state a value in a negotiation has an advantage in terms of the final, agreed value—simply because that stated value acts as an anchor for the other party (Ritov 1996). Given that the goal of science communication is clear and unbiased transfer of information, the ubiquity of anchoring is a problem. When someone elicits your opinion by asking whether you, for example, think that atmospheric CO_2 concentrations will exceed 500 ppm by 2050, then that 500 ppm figure will tend to anchor your subsequent estimate of CO_2 concentration in 2050. Of course, at this point, some people may point out that *estimates* like this are generally not made out of thin air but are, rather, the outputs of complex, detailed modelling. This, however, simply pushes the problem of anchoring back into model selection and design. At some stage, models must make assumptions about possible parameter values that are uncertain (otherwise we would not need models). These values are either estimated by the person making the model or elicited from other experts—and in both cases, anchoring can impact the input values, biasing the outputs.

8.3.3 Up anchor

So, given this, what can be done to limit the impact of anchors on estimates? The key here is understanding the bias's modes of action—as described above. Whether the anchoring bias is resulting from a person deliberately adjusting away from a starting estimate or because of a cognitive priming effect, the key is to break the dependence on a single anchoring value. Thus, in the example above, where you are asked whether CO_2 concentration will reach 500 ppm by 2050, the inclusion of additional comparisons can undermine the influence of any single anchor. That is, after considering whether the concentration will reach 500 ppm, you should also consider the chance of it reaching 450 ppm or dropping to below 400 ppm prior to making your estimate. In this way, the natural tendency of your brain to seek confirming evidence can be put to use by searching for ways in which *any* of these values could be true, which may help in coming to an estimate that is not biased by one, specific value.

Similarly, if you recognise that an anchoring value may be affecting your estimate, then engaging in a process such as the 'consider the opposite' one suggested by Mussweiler *et al* (2000) is probably a good idea. The idea behind this is that, after making your first estimate, you should identify any anchoring value and then use this to generate alternative values—by, for example, considering values that lie on the opposite side of the anchor to your initial estimates and then deciding whether this is feasible. As in the multiple anchor approach above, the goal here is to use this iterative process to generate more options that need to be considered and which will, as a result, weaken the effect of the single, anchoring value.

8.4 Heard it all before?

On reading the above, it is possible that you are experiencing a sense of déjà vu. That may be because the various biases discussed in this book are related in a complex, non-linear way that has, unavoidably, resulted in a book structure that mentions the same biases in different contexts. It may, however, simply be the fact that your memory is fallible—not only do you not remember everything that has happened to you, but you can and do remember things that have not happened.

Déjà vu is an example of this—where aspects of a new situation mistakenly trigger a strong feeling of recognition, leading to the conclusion that you are have experienced the event before. There are, however, even clearer demonstrations of the fallibility of our recall and, of these, perhaps the most important for us is hindsight bias, discussed below.

8.4.1 Mental break

Before talking about hindsight bias, however, we are going to take a short break. Think back to the last time you went to the beach. Try to recall the weather, where exactly you were and something that you did: playing volleyball, lying on a towel, looking in rockpools or swimming? Can you remember?

Now, think about what you are recalling. Do you have an image in your head? Is it an accurate memory of what was happening?

If you are like most people, the answer is probably not. In fact, typically, when asked to reconstruct a personal memory of this sort, the memories differ from reality in one key way— you were probably 'recalling' your actions from the point of view of an external observer. That is, you may have had an image of yourself playing volleyball, swimming or lazing on the beach—but not from the perspective that you would actually have had. Instead, you have imagined what you might have looked like to a third party at the same location rather than recalled your own experience from the event.

This is a fundamental aspect of memory—that it is a reconstructive process rather than an accurate recording of events. While we may remember key aspects of an event, we rarely commit enough detailed to memory to reconstruct it very accurately —which explains, amongst other things, the very poor performance of people in eye-witness identification tasks (for a brief overview of such effects see, e.g. Loftus 1997).

8.4.2 Hindsight bias

A central effect arising from this fact is hindsight bias—our tendency to revise our memories when we come into possession of new information. This results in our believing that we actually knew what we now know all along—or that the outcome of a now observed event was easily predictable in advance.

This was demonstrated by Fischhoff and Beyth (1975) amongst others, who asked people to estimate the likelihood of various outcomes of events and then, after the event had occurred, asked people to recall the probabilities that they had assigned to those outcomes. For example, just prior to President Richard Nixon's 1972 trip to China, people were asked to rate on a 0%–100% scale how likely fifteen different possible outcomes of the visit were. They were then asked to recall these probabilities either a short or long time after the trip had been completed. Finally, they were asked whether they recalled the specific event having occurred or not. The researcher's prediction was that people's 'recollection' of the probabilities they had assigned would be affected by whether the event had occurred (specifically, whether they knew it had occurred) and this was exactly what was found. Two-thirds of people asked two weeks after the event 'remembered' having assigned higher probabilities to events that had happened and lower probabilities to those that had not. This increased to 84% of people when the length of time between prediction and recollection was increased to three or six months.

This effect is typically explained in terms of causal explanations (see, e.g. Nestler *et al* 2008). That is, in trying to make sense of the world, people construct causal stories about how events affect each other and the strength of the causal argument is linked to how likely someone is to believe something to be true. When predicting future events, we are missing the outcome and so need to consider possible causal stories that both lead to and contradict the expected outcome. By comparison, once the outcome is known, it is a much easier task to look for the antecedent causes and thus construct a stronger story. Once this causal story is known, it is difficult for us to see how events could have turned out differently. This means that, when asked to reconstruct our memories of estimates made prior to the outcome, this strong, causal information is included and we overestimate how predictable the event was and, so, how likely we *must* have adjudged it.

8.4.2.1 *I hypothesised that!*

While any effect that results in people misremembering their past estimates is important for the communication of science, hindsight bias is, perhaps, of greater import for the conduct of science and for understanding the basis of people's biases.

Taking the second point, first, recall (accurately, if possible) the discussion of confidence and overconfidence in chapter 7. The fact that people tend to be overconfident hinges on people misremembering how often their predictions are right and hindsight bias explains why this is, regularly, the case. That is, it seems entirely possible that our tendency to construct causal explanations and then project them 'back in time' makes us feel like our predictions have been right more often than they actually were. Projecting this confidence into the future, it seems likely that we would, as a result, predict that we will be right in the future more often than we actually will be—which is the very definition of overconfidence.

The central role of causal stories in hindsight bias is of particular concern for science as the role of a scientist is to examine data and construct hypotheses (causal stories) for how they have come about. While the scientific process dictates that hypotheses be generated prior to data being collected, this is, of course, not always the case. Sometimes it is an accumulation of data that prompts a new hypothesis and, at other times, the data point so strongly in a particular direction that a post hoc hypothesis suddenly seems like the obvious explanation. That is, just as is the case for other causal stories, knowing the outcome can bias a scientist's belief in their hypotheses and makes it harder for them to construct or accept alternative causal explanations. It is possible, in fact, that the very characteristics that define scientists may make them more (rather than less) susceptible to this effect. For example, a trait called 'need for cognition' (NFC; Cacioppo and Petty 1982), which measures a person's self-rated preference for engaging in effortful thought (which seems likely to be higher than typical amongst scientists) has been demonstrated to increase people's tendency towards false memory creation, of which hindsight bias is a subset (Petty *et al* 2009). This makes sense if one considers the tendency of people with high NFC to examine and re-examine their thoughts and memories, blurring the line between what they thought in the past and what they believe now.

To guard against the effect of such memory biases is difficult. After all, if my memory is faulty, how would I know? The answer, of course, as unsexy as it may be, is better book-keeping. If I have written down, in advance, my predictions, any hypothesised causal structures and how confident I am that these will be supported then, after the outcomes becomes known, I have an unbiased baseline against which to compare my current beliefs—that is, those informed by knowing the outcome. This approach is good practice and forms the basis of the move in various areas of science towards pre-registration of experiments. That is, putting on the record your experimental design and hypotheses so that you and those looking at your papers know that your beliefs are not simply the result of hindsight bias.

8.5 Conclusions

That this chapter occupies a central place within this book is appropriate. While, the book is, thematically, divided into different types of biases, memory is a fundamental part of any decision-making process and, as a result, errors in memory are central to the production to many biases.

All of this results from the fact that memory is not a video camera capturing everything we see but rather an active process wherein our conscious attention and various contextual factors determine what is and is not stored (and how well it is stored). This then requires that remembering be a reconstructive process—where our current context determines how easily or accurately different memories can be retrieved—and, as a result, it is difficult to keep current knowledge distinct from our memories.

This means that the fallibility of our memory can result in biases in: which information we use to make estimates (anchoring); how likely we believe different events are (availability); how often we believe our predictions have been right (overconfidence); and what we believed at different points in the past (hindsight bias). Understanding this is, thus, essential to understanding the ways in which scientists can limit the impact of these biases on both their own behaviour and within their communication.

References

Anderson J R and Schooler L J 1991 Reflections of the environment in memory *Psychol. Sci.* **2** 396–408

Atkinson R C and Shiffrin R M 1968 Human memory: a proposed system and its control processes *Psychol. Learn. Motiv.* **2** 89–195

Australian Bureau of Statistics 2016 *3303.0—Causes of Death, Australia, 2016* (Canberra: Australian Bureau of Statistic) www.abs.gov.au/ausstats/abs@.nsf/mf/3303.0

Baddeley A D and Hitch G 1974 Working memory *Psychol. Learn. Motiv.* **8** 47–89

Bureau of Infrastructure, Transport and Regional Economics 2016 *Road Deaths Australia, Dec 2016* (Canberra: Department of Infrastructure and Regional Development) https://bitre.gov.au/publications/ongoing/road_deaths_australia_monthly_bulletins.aspx

Cacioppo J T and Petty R E 1982 The need for cognition *J. Person. Social Psychol.* **42** 116–31

Chapman G B and Johnson E J 1999 Anchoring, activation, and the construction of values *Organ. Behav. Hum. Decis. Process.* **79** 115–53

Ferguson C J and Heene M 2012 A vast graveyard of undead theories: publication bias and psychological science's aversion to the null *Persp. Psychol. Sci.* **7** 555–61

Fischhoff B and Beyth R 1975 I knew it would happen: remembered probabilities of once-future things *Organ. Behav. Hum. Perform.* **13** 1–16

Fischhoff B, Slovic P and Lichtenstein S 1978 Fault trees: sensitivity of estimated failure probabilities to problem representation *J. Exp. Psychol. Hum. Percep. Perform.* **4** 330

Furnham A and Boo H C 2011 A literature review of the anchoring effect *J. Socio-Econ.* **40** 35–42

Hamann S 2001 Cognitive and neural mechanisms of emotional memory *Trends Cogn. Sci.* **5** 394–400

Loftus E F 1997 Creating false memories *Sci. Am.* **277** 70–5

Marshall P H and Werder P R 1972 The effects of the elimination of rehearsal on primacy and recency *J. Verb. Learn. Verb. Behav.* **11** 649–53

Miller G A 1956 The magical number seven, plus or minus two: some limits on our capacity for processing information *Psychol. Rev.* **63** 81

Mussweiler T, Strack F and Pfeiffer T 2000 Overcoming the inevitable anchoring effect: considering the opposite compensates for selective accessibility *Person. Social Psychol. Bull.* **26** 1142–50

Nestler S, Blank H and von Collani G 2008 Hindsight bias and causal attribution: a causal model theory of creeping determinism *Social Psychol.* **39** 182–8

Petty R E, Briñol P, Loersch C and McCaslin M J 2009 The need for cognition *Handbook of Individual Differences in Social Behavior* (New York: Guilford) pp 318–29

Ritov I 1996 Anchoring in simulated competitive market negotiation *Organ. Behav. Hum. Decis. Process.* **67** 16–25

Slovic P, Fischhoff B, Lichtenstein S and Roe F J C 1981 Perceived risk: psychological factors and social implications *Proc. R. Soc. Lond.* A **376** 17–34

Taronga Conservation Society Australia 2016 *Australian Shark Attack File Annual Report Summary for 2016* (Sydney: Taronga Conservation Society Australia) https://taronga.org.au/conservation/conservation-science-research/australian-shark-attack-file/2016

Tulving E 1972 Episodic and semantic memory *Organ. Mem.* **1** 381–403

Tversky A and Kahneman D 1973 Availability: a heuristic for judging frequency and probability *Cogn. Psychol* **5** 207–32

Welsh M B, Begg S H and Bratvold R B 2006 SPE 102188: correcting common errors in probabilistic evaluations: efficacy of debiasing *Society of Petroleum Engineers 82nd Annual Technical Conf. and Exhibition (Dallas, TX)*

Welsh M B, Rees N, Ringwood H and Begg S S H 2010 The planning fallacy in oil and gas decision-making *APPEA J.* **50** 389–402

Welsh M, Steacy S, Begg S and Navarro D 2016 A tale of two disasters: biases in risk communication *38th Annual Meeting of the Cognitive Science Society (Philadelphia, PA)* ed A Papafragou, D Grodner, D Mirman and J Trueswell (Austin, TX: Cognitive Science Society)

Wilson T D, Houston C E, Etling K M and Brekke N 1996 A new look at anchoring effects: basic anchoring and its antecedents *J. Exp. Psychol.* **125** 387

IOP Publishing

Bias in Science and Communication
A field guide
Matthew Welsh

Chapter 9

Angels and demons: biases from categorisation and fluency

This chapter focuses on another, central cognitive process: categorisation. That is, how people sort objects, events and even other people into types or groups in order to better understand or predict the world. This is, of course, necessary because our memories are limited (as described in the previous chapter), with the result that we cannot remember each individual event or object in all of its glorious detail but must, instead, often rely on similarities between events or sketches of 'typical' behaviours in order to predict outcomes and actions.

While necessary, any attempt at categorisation involves a loss of information and, the broader the category used, the worse its predictive power will tend to be. As a result, categorisation leads to biases in people's behaviours and understanding as they use imprecise assumptions based on category membership to predict the behaviour of individuals. This underlies a number of known biases, including the halo effect and stereotyping, discussed below.

9.1 Heights of success

Imagine that you are selecting your favoured local representative in an election. What traits do you look for? For most people, one imagines, this would include things like the candidate's intelligence, ethics, political savvy and their ability to negotiate for their electorate. In addition, you may include additional characteristics dependent on your political ideals. For example, a neo-liberal may want someone who has already found success in business before moving into the public sphere.

Or, you could do what many people do and just select the taller candidate.

The above is, of course, an oversimplification but there is strong evidence, based on over a century of research, that height is a predictor of election success—at least in elections such as the US Presidential ones, where two candidates are directly compared (for a recent review, see, Stulp *et al* 2013). Stulp *et al*'s detailed analysis indicates that, across the 42 US Presidential elections for which candidate height data existed, the taller presidential candidate won the popular vote 67% of the time (although the electoral college system translated this to only a 58% chance of winning the presidency) and that the difference in height between the candidates correlates at 0.365 with the proportion of the popular vote received. Simply put, tall candidates get more votes. The question, of course, is why?

9.1.1 The halo effect

The answer to this question seems to be related to a bias known as the 'halo effect'. This refers to our tendency to assume that possession of one positive characteristic means that people will have other, positive characteristics—or, even, that we will interpret neutral characteristics positively in cases when the person displays *one* positive characteristic. For example, Nisbett and Wilson (1977) presented people with two video recordings of the same man. In one, the man was warm and friendly while, in the other, he was cold and distant. That people liked the man in one and disliked him in the other was, as a result, not surprising. What surprises most people, however, is that, when asked to assess the man's other (unchanging) characteristic like his appearance and accent, people rated these as either increasing or decreasing their liking of the man in line with their overall impression. That is, people shown the warm and friendly video found the man's accent and appearance appealing, while those shown the cold and distant video found those same characteristics irritating.

This is the halo effect—which seems to reflect our expectation that good (or bad) traits will tend to cluster together. Or, put another way, that people are easily sorted into 'good' or 'bad' and, thus, knowing just one of their traits is sufficient to accurately categorise them.

This belief overrides our ability to objectively judge someone's characteristics. As a result, the behaviour and characteristics of people whom we like tend to be judged far more leniently than the same behaviours and characteristics as displayed by people we dislike.

9.1.2 Pretty/good

So, how does this explain the success of tall politicians? The answer is that height is a key marker of health—simply because good nutrition allows a person to grow taller, stronger and smarter. While not always particularly strong correlations, these are all observable tendencies. Tall people *are* (on average) smarter than shorter people (see, e.g. Lynn 1989, Pearce *et al* 2005). Height is, similarly, regarded as a component of attractiveness, with taller men, in particular, being regarded as better matches—a point driven home by Cameron *et al*'s (1977) observation that 80% of US women's dating profiles/advertisements indicated a desired height of 6 feet or more, which a comparison with the actual heights of American men reveals limits their choices to the tallest ~25% of men. Finally, tall people tend to be more successful—having a greater chance of being employed and earning more money (although whether this is a separate effect or a consequence of greater intelligence has been disputed, see. e.g. Case *et al* 2008).

So, simply by being taller, a person is associated with a cluster of related, positive characteristics—being healthy, wealthy and wise—and the halo effect can thus result in us interpreting their other, neutral characteristics more positively as well. When it comes time to select a leader, then, tall people will tend to be viewed more positively overall—explaining their greater success at the polls.

One question that the above may prompt is: if these traits are positively correlated, is relying on them the wrong thing to do or should we be happy to hire the taller person for a job or elect them as our representative? That is, if the tall person is likely to be smarter than their opponent, why not just choose them? The problem with this, of course, is that the weak correlations (~0.2 between height and intelligence) are measured across the whole population whereas, when selecting candidates for a job, these candidates have already *been* selected. That is, the predictive power of the obvious trait (height) will have been reduced by range truncation—as shown in figure 9.1.

Looking at the figure, one sees the danger of relying on a correlation derived from a general population for predicting the performance of a selected population. Here, the top 10% of the IQ distribution has been selected—that is, people with IQs of 120 or above (typical of people with higher degrees like MDs or PhDs). Within this, truncated, sample there is effectively *no* correlation between height and IQ, despite the overall trend seen in the larger population. This is, however, unlikely to affect people's expectations regarding the relationship between height and competence,

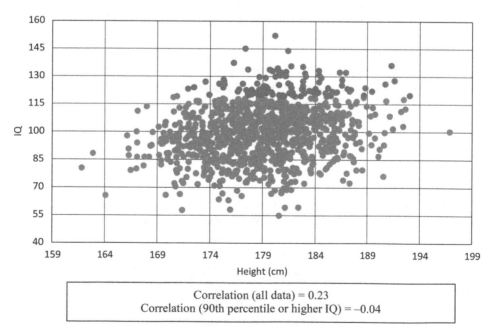

Correlation (all data) = 0.23
Correlation (90th percentile or higher IQ) = −0.04

Figure 9.1. Demonstration of range truncation in height/IQ relationship. Note: gridlines represent intervals of 1 SD centred on the means of 100 (IQ) and 179 (height).

9-3

with the result that a taller candidate may still be hired preferentially—simply because people tend to rely on these broad cues rather than considering the specifics.

Such associations affect people's judgements in a wide range of situations beyond hiring and elections, including some that are specific to the pursuit and communication of science. For example, we often assume—with some justification—that scientists working at high-prestige institutions are more competent than those working at lower prestige institutions. The result of this assumption, however, is likely to be a halo effect where the work that those scientists do will be regarded as superior to the same work presented by a scientist with a less prestigious affiliation, simply because one positive characteristic tends to beget more positive appraisals.

9.1.3 Him that hath

Peters and Ceci's (1982) study offered some support for this position: changing the author name and institution of papers published in high-prestige outlets by psychologists from high-standing universities resulted in the majority being rejected for 'serious methodological flaws' by those same outlets, despite the papers not being recognised as resubmissions. While subject to various methodological concerns (see, e.g. the commentaries following Peters and Ceci's article, which suggest—amongst other possible causes—the possibility that the result was caused by peer-review in psychology being effectively random rather than biased), that this might be true is not, particularly, surprising. The 'Matthew effect' (Merton 1968, 1988) describes the tendency for acclaim to attract further acclaim—at the expense of other equally or more deserving work. This is named for the passage from the Gospel of Matthew: *For whosoever hath, to him shall be given, and he shall have abundance: but whosoever hath not, from him shall be taken away even that which he hath.*

That is, once a person (or institution) becomes recognisable for quality work, their other work is assumed to be of high quality by association and the division of credit between high- and low-prestige collaborators strongly favours the person with higher starting prestige. This, it seems, harks back to the halo effect and the way in which we categorise people. It is easier for us to remember a key figure and attribute all of the relevant work in a field to them rather than trying to understand the complex tangle of multiple people's contributions to a topic. This can result in senior academics receiving credit for work predominantly done by their junior colleagues and even having work they did not contribute to at all being misattributed to them (Merton 1968).

9.2 In and out

A commonly recognised extension of this idea of categorisation underlying bias is stereotypical thinking—that is, the tendency we have to sort people into categories based on broad categories such as 'race', 'gender' or 'ethnicity' and use those category memberships to make decisions about how a person will behave. This produces a variety of effects, including distinctions drawn between members of our own 'in-group' (i.e. people like us) and members of 'out-groups' (i.e. anyone else) that interact with various of the biases discussed throughout this book to produce biases in our decision making.

9.2.1 Leading lights

Before going on, stop and complete the following task:
1. List five scientists from your field of expertise/interest.
2. List five famous scientists (from any field or fields).

9.2.2 The stereotypical scientist

What are the attributes of a stereotypical scientist? Think about scientists you know or know of and what they have common. Which of these traits seem central—the essence of being a scientist? This is your stereotypical scientist.

In fact, of course, what is counted as stereotypical will depend who you ask. To different people, descriptions of 'scientists' might range from 'objective seekers of truth' to 'bubble-dwelling, leftist elites using their alleged expertise to try to control society'. Within a particular group, however, there may be a set of commonly used descriptors that one could regard as the core of a stereotype. In fact, this was how stereotyping was first measured—college students were asked to describe the attributes most typical of particular ethnic groups and the most commonly chosen of these formed the 'stereotype' (Katz and Braly 1933).

When one hears the word 'stereotype', it almost always has negative connotations —that is, it carries with it an implication of falsehood or bias. In fact, early work defined stereotypical beliefs as inherently wrong—based on the fact that group membership is never perfectly aligned with possession of any set of traits. That is, because surprisingly few members of a defined group actually possess all of the stereotypical attributes of that group, it seems that stereotypical views are, generally, wrong. That, however, prompts a question: if stereotypes are wrong, why do people have stereotypical beliefs?

The simplest answer seems to be that the above is too simple a representation of how people understand group membership. Work on categorisation has demonstrated that categories are rarely sharply defined. Rather, they have fuzzy boundaries that allow for individuals to have only a subset of features while still being regarded as members of a particular type (for example, people are perfectly happy to accept that the extinct moa from New Zealand were birds, despite their complete lack of a stereotypical bird feature—wings).

This results in memberships being probabilistic rather than deterministic (for a discussion, see, e.g. Judd and Park 1993). That is, rather than the specific traits of individuals being used to definitively rule them into or out of groups, the general characteristics of groups are used to decide—on a probabilistic basis—whether a new individual will possess a particular trait. For example, the stereotypical lawyer has (let us presume) a large vocabulary. When meeting a new lawyer, then, the stereotype informs us that this person, too, is *likely* to understand erudite or even inelegantly verbose communiques.

Recognition that stereotypes are a type of social category membership with predictive power helps us to understand why people so readily form these impressions. As pointed out earlier in the chapter (as in chapter 8), a key to understanding human cognition is the recognition of our limitations. We do not have infinite mnemonic or processing capacities and so we need to make generalisations about how people will behave based on limited information. Categories are a key way of doing this—compressing knowledge about a large group of events or individuals into a form that can more easily be applied to new instances.

So, stereotypes are not inherently bad. They come about because of our need to predict the world using limited information. Problems arise, though, when other limitations affect these categories, leading us to make biased predictions. For example, the halo effect described above demonstrates a key problem with not taking all of the features of an individual into account; it results in the assumption that sets of related features will co-occur. That is, that possession of one desirable feature means that someone also has other desirable features.

It also results in the sorts of errors that people think of when they hear the word stereotype—when reliance on stereotypes conflicts or interacts with other evidence. Recall, for instance, confirmation bias (discussed in chapter 5), which prompts people to search for information that accords with their already held beliefs *and* to weight confirmatory evidence more heavily than disconfirmatory. The result of this is that someone who holds a stereotype will tend to 'see' more evidence that supports rather than undermines their stereotype. In particular, ambiguous evidence is very easily interpreted in line with a stereotype. For example, Sagar and Scholfield (1980) demonstrated that ambiguously aggressive behaviours (e.g. pushing, poking, taking someone's pencil or asking for someone else's food) were more likely to be rated as threatening rather than playful when perpetrated by black actors rather than white. This was explained as the stereotype of black people being more aggressive affecting the observer's interpretation of the action—in order to reinforce their stereotype.

A combination of stereotypes, the halo effect and confirmation bias gives rise to the situation where a person can have a view of a group that is largely positive or negative (as these traits are assumed to cluster) and will tend to not only look for examples that confirm these opinions but also interpret ambiguous evidence as supporting them as well.

This holds true of most people to a greater or lesser extent—simply because we share the same underlying cognitive processes that give rise to these biases. That is, 'biased' could accurately be described as being the natural state of people. Even people who draw no distinctions based on race or ethnicity often hold strong stereotypical views on other groups—think, for example, about the partisan, political divide and the views that conservatives have regarding progressives and vice versa.

The key thing to remember though, is that stereotypes exist for a reason—they are categories and, as such, designed to enable us to predict behaviour in advance. The larger the group a stereotype covers, however, the less predictive power they are likely to hold. For example, the stereotypical features and behaviours of 'birds' are far less predictive of how a sulphur-crested cockatoo will look and behave than the stereotypical 'parrot' or, at an even finer scale, 'cockatoo'. This means that broad stereotypes' predictive power tends to be weak and should be quickly overridden by evidence regarding any, specific individual. Other cognitive limitations, however, prevent this occurring as quickly as one might expect and, importantly, tend not to affect the underlying the stereotype. To continue the above example, learning how an individual sulphur-crested cockatoo tends to behave is unlikely to affect a person's stereotypical 'bird' much, if at all. Instead, the individual is marked as an exception to the still-standing rule.

9.2.3 Implicit bias

The observation that stereotypes are resistant to change can lead to bias in two distinct ways. First, as above, it can result in us predicting that a person will behave in a particular manner because they belong to a particular group. Equally, however, it could lead us to believe that a person will *not* possess core qualities of the stereotype, which can be equally damaging.

Think back, for example, to the question we considered above: what is a stereotypical scientist? To determine this, you were asked to think about scientists you know and their traits. Given the large gender imbalance in science careers (and the even larger, historical imbalance) it is extremely likely that majority of the scientists you considered were male and thus, unconsciously at least, so was your 'stereotypical' scientist. This is not to say that, when asked to list the *essential* traits of a scientist gender would have been included but it does mean that, in situations where someone's resemblance to this stereotype is important, a man is likely to be adjudged more 'scientist-y' than a woman.

Work on implicit bias supports this view—with research using the implicit association test across 34 countries and more than 500 000 participants demonstrating that, in general, people are faster at completing a categorisation tasks with items that are stereotypically related, such as 'male' and 'science', and relatively slower at categorisation when faced with less related items like 'female' and 'science' (Nosek *et al* 2009). The implications of measures of implicit bias for individuals' actual behaviour is hotly debated (see, e.g. Oswald *et al* (2015) for a recent review of whether implicit bias predicts prejudiced behaviour) but stereotypes seem to be quickly learnt and do affect people's views and behaviour.

For example, Bian *et al*'s (2017) study demonstrated that, by age 6, girls' views of themselves as different from boys in terms of intelligence are well entrenched, with fewer endorsing the view that members of their own gender were likely to be 'really, really smart'. This idea—that brilliance is a gendered trait—seems to come about naturally from exposure to societal evidence. Consider, for example, well-known scientists and science communicators. What proportion of these are female compared to male? Similarly, the poster-children for genius across a variety of fields are almost exclusively male—think: Shakespeare, Einstein, Newton, Michelangelo, Leonardo—in fact, all of the ninja turtles.

As a result, without any need for parents or teachers to directly link scientific (or other) brilliance to masculinity, children learn the category relationships or stereotype. More concerningly, they then begin to act accordingly. Bian *et al*'s (2017) study, for example, also showed 6-year-old girls self-selecting away from tasks that were described as being suited for brilliant children. This effect, carried forward across schooling, is likely contributing to the well-identified gender imbalance in science, technology, engineering and mathematics (STEM). That is, female students, despite equally good grades, tend to self-select away from the 'harder' maths and science subjects in high school, resulting in fewer females pursuing STEM careers. This, combined with the 'leaky pipeline' for females in academic STEM positions, acts to reinforce the stereotype of the scientist as male— completing the circle.

That is, the simple fact that people naturally categorise things into types affects the predictions that they make and thus their behaviour in such a way as to reinforce the stereotype.

9.2.4 You remind me of me

Perhaps the most obvious situation where the sorts of categorisation biases noted above might occur within science is in job interviews. While metrics such as patents granted, papers published, *h*-indices and or student evaluations are typically used to develop a short-list of candidates for a position, selection often comes down to an interview process and it is here that stereotypes and halo effects are most likely to have an effect.

To some extent, for example, the interviewee is being assessed against a stereotype —of a good scientist, researcher or lecturer, for example, which may advantage men applying for such positions as these are, more typically, male-dominated areas. Even without so clear a bias, however, one needs to consider that all stereotypes are fuzzy constructs. That is, there are a variety of ways in which one could be, for example, a good researcher—and candidates may not be directly comparable in terms of their good and bad features. At this stage, therefore, some subjective judgement is required. For example, is an applicant for a post-doctoral position with two first authored papers as good as one who has made a lesser contribution to six papers during their PhD candidature? This may, in fact, depend on the laboratory in which they were working and the other people therein—making it very difficult for an outsider to judge which cues are valid determinants of candidate quality.

What then, are the cues available to decide whether candidate A or candidate B will be a better researcher? A concern here is that a researcher on the interview panel, rather than making this complex decision, makes a simpler one, asking: how similar to me is this candidate? After all, they reason (unconsciously, one presumes): I am a good researcher—otherwise I would not be here (a claim strengthened in their mind by the action of overconfidence as discussed in chapter 7)—so people like me are likely to be good researchers too.

A key problem here, of course, is that the category 'like me' can be assessed at a very superficial level and comes with halo effects. For instance, if a person holds stereotypical views about certain ethnic groups, then a likely outcome is that they will judge people belonging to the same ethnic group as them as possessing more positive traits—being smarter, nicer, more reliable and so forth—even when there is no evidence of those traits. The more similar to the interviewer the interviewee is—whether on ethnic, gender or social grounds—the more one expects positive assessments of those less tangible traits may make the difference between being hired or not (see, e.g. Howard and Ferris 1996).

This tendency towards homogeneity, in addition to resulting in situations where equally or even better qualified candidates are overlooked for superficial reasons, produces worse outcomes for the hiring company or university as a lack of diversity of views produces measurable detriments in a variety of situations.

9.2.4.1 Get with the program

A core example of the problems caused by a lack of diversity is the effect known as Groupthink—the tendency of groups to come to an overconfident, 'consensus' view (Janis 1971). This results in a group being convinced of its own expertise, suppressing internal dissent and producing decisions that are markedly overconfident compared to individuals and, importantly, occurs more swiftly when groups are composed of more similar individuals and has less effect when groups are more diverse. A group that is low in diversity is also likely to produce fewer novel ideas—a significant concern for the pursuit of science.

9.3 Easy to believe

Reading the above—or any other argument put forward herein—I would hope that the logic of my argument and the strength of the supporting documentation will convince you that the positions I am putting forward are true. However, I am also aware that whether you believe me or not is dependent on a number of other factors. For example, as noted above, whether you have heard of me personally or of my institution may well bias your regard for my argument (whether positively or negatively). Another factor, however, is how easy my argument is to follow.

If my examples are simple and my argument clear, you are, of course, more likely to understand them. What may seem more surprising is that, the simpler and clearer my exposition is, the more likely you are to also *believe* me. That is, accept my position as true.

This results from fluency, which refers to the sense of ease (or, conversely, difficulty) with which we perform any particular cognitive processing. If something comes easily, we are more inclined to believe it—which makes sense in some circumstances. For instance, if I say '2 + 2 = 4', you are likely to have heard that many times before, reducing the processing required to understand what I am saying and resulting in an immediate feeling of fluency. That is, feelings of fluency accompany oft-repeated facts—making us confident of their truth without us actually having to double-check.

Fluency is, however, affected by factors other than repetition. In fact, *anything* that makes a task cognitively harder reduces the feeling of fluency while things that make it easier increase that feeling. To take a surprising example, Reber and Schwarz (1999) asked people to decide whether a series of statements—of the form 'town A is in country B'—were true or not. All of the statements faded into view over time, but some participants were shown these statements in dark, easy to read colours on a white background, while other were shown the same questions but in paler, harder to read colours (e.g. yellow against a white background). By the end of the 'fade-in' all were legible and the fact that participants found the paler colours harder to read—as judged by response times—is probably unsurprisingly. They were also, however, more likely to endorse these visually clearer statements as true—by a small but statistically significant margin. That is, simply making a statement easier to read makes people more inclined to believe its truth.

Such effects have been shown across a wide variety of fields (for a review, see Oppenheimer 2008) and have even been used as a method for reducing decision-making bias. To understand how this might work, recall the distinction drawn back in chapter 2 regarding the two systems approaches to decision making. In this approach, simple intuitive processes (system 1) automatically activate to generate possible answers to questions and it is only when these fail or when our error checking mechanisms flag a response for some reason that we activate our rational (system 2) reasoning.

Adding fluency to this story, we find ourselves in a situation where intuitive responses that are easily (fluently) generated are more likely to be believed to be true and thus not checked. The corollary of this, of course, is that, if you make the task feel less fluent, people will be less sure of their responses and more inclined to check them. This is exactly what Alter *et al* (2007) found—demonstrating that people's performance on the cognitive reflection test (a series of three questions with strongly intuitive, but wrong, responses; Frederick 2005) could be improved by presenting the test to them in a distorted font. This increased the processing required to read the questions, making their overall test performance feel less fluent and thus prompting them to check their answers and discover their errors more often.

9.3.1 Caveats and clauses

This is, of course, of central importance for the communication of science (and the propagation of factoids and myths as will be discussed at length in chapter 13). Here, the complex, caveat-ridden, multi-clause communications adopted by scientists—designed to increase the precision of discussions amongst themselves—are likely to decrease the sense of fluency of anyone else hearing the argument. This, in turn, decreases the likelihood that the listener will believe the message—particularly if it is competing with a simple, oft-repeated catch-phrase.

Given this, as pre-empted in chapter 3, scientists need to consider the *aim* of their communication. Is it to explain all of the details of a situation to someone, in order to allow them to make up their own minds through a process of critical reasoning? Or is it to convince listeners of widely accepted facts? Each of these requires a different communication strategy but, in all cases, a polished, easy-to-read statement will increase the target's sense of fluency and thus how likely they are to believe you.

9.4 Conclusions

The above discussions highlight a core reason for this book's existence. A variety of effects with significant effects for not just science and scientists but wider society can be traced back to limitations in human cognitive capabilities. Peculiarities in basic mental processes like categorisation predictably give rise to—and reinforce—a wide array of biases that affect our judgements about important things like who we should vote for, who we should hire and whether or not we believe what we are told.

If we want to avoid these biases, we need to understand how and under which conditions they arise. Only with this knowledge can we gain insight into how our own and others behaviour are being affected by stereotypes and related effects and then take steps towards removing their influence on our behaviour through conscious counteraction as described in later chapters (part 3).

References

Alter A L, Oppenheimer D M, Epley N and Eyre R N 2007 Overcoming intuition: metacognitive difficulty activates analytic reasoning *J. Exp. Psychol.* **136** 569

Bian L, Leslie S J and Cimpian A 2017 Gender stereotypes about intellectual ability emerge early and influence children's interests *Science* **355** 389–91

Case A and Paxson C 2008 Stature and status: height, ability, and labor market outcomes *J. Polit. Econ.* **116** 499–532

Cameron C, Oskamp S and Sparks W 1977 Courtship American style: newspaper ads *Fam. Coord.* **26** 27–30

Frederick S 2005 Cognitive reflection and decision making *J. Econ. Persp.* **19** 25-42

Howard J L and Ferris G R 1996 The employment interview context: social and situational influences on interviewer decisions *J. Appl. Social Psychol.* **26** 112–36

Janis I L 1971 Groupthink *Psychol. Today* **5** 43–6

Judd C M and Park B 1993 Definition and assessment of accuracy in social stereotypes *Psychol. Rev.* **100** 109

Katz D and Braly K 1933 Racial stereotypes of one hundred college students *J. Abnorm. Social Psychol.* **28** 280

Knobloch-Westerwick S, Glynn C J and Huge M 2013 The Matilda effect in science communication: an experiment on gender bias in publication quality perceptions and collaboration interest *Sci. Commun.* **35** 603–25

Merton R K 1968 The Matthew effect in science *Science* **159** 56–63

Merton R K 1988 The Matthew effect in science, II: cumulative advantage and the symbolism of intellectual property *ISIS* **79** 606–23

Nisbett R E and Wilson T D 1977 The halo effect: evidence for unconscious alteration of judgments *J. Person. Social Psychol.* **35** 250

Nosek B A *et al* 2009 National differences in gender–science stereotypes predict national sex differences in science and math achievement *Proc. Natl Acad. Sci.* **106** 10593–7

Oppenheimer D M 2008 The secret life of fluency *Trends Cogn. Sci.* **12** 237–41

Oswald F L, Mitchell G, Blanton H, Jaccard J and Tetlock P E 2015 Using the IAT to predict ethnic and racial discrimination: small effect sizes of unknown societal significance *J. Person. Social Psychol.* **108** 562–71

Pearce M S, Deary I J, Young A H and Parker L 2005 Growth in early life and childhood IQ at age 11 years: the Newcastle Thousand Families Study *Int. J. Epidemiol.* **34** 673–7

Peters D P and Ceci S J 1982 Peer-review practices of psychological journals: the fate of published articles, submitted again *Behav. Brain Sci.* **5** 187–255

Reber R and Schwarz N 1999 Effects of perceptual fluency on judgments of truth *Consc. Cogn.* **8** 338–42

Sagar H A and Schofield J W 1980 Racial and behavioral cues in black and white children's perceptions of ambiguously aggressive acts *J. Person. Social Psychol.* **39** 590

Stulp G, Buunk A P, Verhulst S and Pollet T V 2013 Tall claims? Sense and nonsense about the importance of height of US presidents *Leadersh. Quart.* **24** 159–71

IOP Publishing

Bias in Science and Communication
A field guide
Matthew Welsh

Chapter 10

Us and them: scientists versus lay-people and individual differences in decision bias

Up until now, we have focused on biases that result from general, human tendencies in decision making. While this is a valuable approach, allowing us to talk about how people typically respond in given situations, we also need to consider the ways in which that people's abilities differ. For example, if a bias is caused by limitations in memory, should we expect a person with a better memory to be less affected?

This chapter considers such differences between people—both in terms of the attributes of different groups of people and in terms of the natural variation in cognitive abilities and tendencies within any population—and how this influences susceptibility to various biases.

10.1 Experts and novices

On hearing about biases, one of the first questions that many people have is: are experts less susceptible or immune to biases? That is, does knowing more about some field of expertise prevent biases affecting a person's judgement? This is a sensible question, particularly given the observation from previous chapters that biases are often linked to uncertainty and an expert should have *less* uncertainty—although, as noted in chapter 3, the situations where expertise is most needed are often those where uncertainty is highest

On the side of expertise aiding decision making, we have significant evidence of expertise producing better outcomes. In fact, that is largely how we define expertise— the ability to perform some task or tasks better than a novice could—and researchers looking at expert decision making in the field have demonstrated that this is the case. (Defining expertise is, in fact, quite difficult in the absence of a set of tasks designed to differentiate between differing levels of expertise and people often simply use 'experience' as a proxy—see, e.g. Malhotra *et al* 2007.)

For example, the field of naturalistic decision making (NDM) examines expert decision making in environments with time pressure and high consequences—such as military command and firefighting (see, e.g. Zsambok and Klein 1997). The primary findings of this field have been that: experts in these environments do, typically, produce good outcomes; their solutions to problems come to them intuitively; and that these intuitions come about through the decision makers' *situational awareness*—that is, their understanding of the different types of decisions that they need to make and appropriate strategies for each, learnt over repeated exposures to similar situations.

The question, then, is whether this evidence from NDM research holds for other areas of decision making—and the overall pattern of results from the decision-making literature suggests that expertise tends *not* to provide significant protection against biases when, for example, experts are asked to provide estimates of uncertain parameters.

Recall, for instance, the Welsh *et al* (2010) paper noted in section 8.2.4.2 as part of the discussion of the effect of availability on the planning fallacy. In this experiment, experts were shown to be *more* affected by the unpacking of a general category into a set of subcategories than were the student comparison groups.

Additionally, in line with the explanation of anchoring shown in figure 8.2, while expertise within an area can limit the values that a person considers feasible and thus the extent to which an anchor can affect a person's estimate (see, e.g. Welsh *et al* 2014), experts are still affected by anchoring values. The result of this is that, under conditions of high uncertainty, anchors have a significant effect on expert estimates. Work by Englich and colleagues (Englich and Mussweiler 2001, Englich *et al* 2005, 2006), for example, demonstrated that judges sentencing decisions are affected by anchoring values—whether provided by the prosecution's demands for long sentences or even numbers randomly determined in front of the judge—and expertise or experience does not change this.

Similarly, a variety of researchers have examined the impact of expertise on overconfidence (specifically, overprecision as described in chapter 7). Examination of the ranges provided by experts and novices in a variety of fields demonstrates that expert ranges are, as one would expect, more accurate. That is, they are centred closer to the true value. However, the experts' ranges are also markedly narrower and thus contain the true value no more often than the novices (see, e.g. McKenzie *et al* 2008). That is, the benefit provided by expert's greater knowledge tends to be balanced by their increased confidence with the result that their level of over-confidence remains very similar to that of novices.

An exception to this, however, is the performance of meteorologists, who display very good calibration on their forecasts of the likelihood of rain (see, e.g. Murphy and Winkler 1984). Why meteorologists differ from other experts is an important question to answer and forms the foundation of a key discussion between the leading lights of the naturalistic and heuristics and biases approaches to decision making regarding the role of expertise in decision making.

10.1.1 Trust me, I am an expert

In this paper, Kahneman and Klein (2009) discuss the conditions under which you should trust expert judgement—drawing on their individual research programs that have demonstrated very poor and very good outcomes, respectively, when experts rely on intuition. This is relevant to the example of meteorologists given above because the primary distinction that Kahneman and Klein agreed upon was the information structure of the environment in which the experts were operating.

Specifically, if an expert operated in an environment where they saw large numbers of exemplar problems, made decisions and then received fast and accurate feedback on how well their implemented plan worked, then you could trust that, after some duration, they would become experts in that area and their intuitions could safely be assumed to be driven by their automatized situational awareness. This is what is seen with meteorologists' predictions of rain: they make predictions every day; and receive accurate feedback on the quality of these predictions the following day (or week). Similarly, a fire-fighter experiences fires on most days and receives immediate feedback on the quality of their decisions about how to fight a particular fire.

In many fields, however, timely and accurate feedback is in short supply—or may never arrive at all. For instance, a geologist asked to estimate the *average* porosity of an oil reservoir will *never* find out what the true average porosity of that rock is. Continued exploration, development and production of oil may narrow the possible range down, but the average will never be found because doing so would require sampling every point within the reservoir rather than just the locations where wells are located. Even for those estimates that could, in theory, become known—like predictions of the oil price for when a new well is going to come online—are rendered useless from a learning perspective by the time lag between predictions and feedback. This is because, in many industries and fields where expert judgement is required (e.g. oil exploration, pharmaceuticals, engineering design or defence decisions) the delay between a prediction being made and the outcome actually becoming known can be months, years or even decades—with the result that the people who provide an estimate may not even be in the same position or company to receive the feedback, let alone learn from it, when it arrives.

Thus, the conclusion that must be reached is that, in many areas, an expert's intuition is unlikely to be based on solid learning from feedback and, as a result, cannot be assumed to reflect genuine expertise.

10.2 Individual differences

If expertise (or its poor cousin, 'experience') is not a reliable indicator of resistance to bias, then the obvious next step is to consider traits that differ across the general population and whether any of these predict immunity or lessened susceptibility to biases. Work in psychometrics—the area of psychology focused on individual differences in abilities or tendencies—has provided a wide variety of possible traits, as detailed below. It should, however, be noted that the majority of work on heuristics and biases has been conducted under the banner of 'cognitive psychology'—the principle aim of which has always been to seek general tendencies rather than individual differences.

This division within scientific psychology is long-standing (Deary 2001 for instance cites nine previous papers dating back to 1957 calling for this gap to be bridged) and had resulted in individual differences in decision making being a relatively small and less-well explored part of the overall literature. As a result, some of the below is, necessarily, patchier and more speculative in nature.

10.2.1 Memory

While aspects of memory actually form part of the measures of intelligence described below, it is included separately here due to its apparent centrality in a number of biases (discussed in chapter 8). As noted previously, memory is central to the explanation of the availability of heuristic and accompanying biases but, as yet, there seem to have been no studies directly measuring individual differences in short term memory capacity, retrieval fluency or strength of context dependence of memory for comparison with susceptibility to biases resulting from availability.

In contrast, attempts have been made to explain overprecision (overconfidence in the form of too narrow predicted ranges; see section 7.2.4) in terms of memory limitations. For example, the decision-by-sampling approach (Stewart *et al* 2006) and the naïve sampling model (Juslin *et al* 2007) accounts of overprecision both link the production of narrow ranges to human memory limitations. Specifically, arguing that we produce too narrow ranges because any sample of instances we draw from memory is limited by out short term memory (STM) capacity and will thus be small and tend to underestimate the range of possible values in our set of memories. These accounts did not, however, include direct measurement of STM capacity—which is important as this is known to (consistently) vary between five and nine items in different individuals (Miller 1956).

Direct tests of these predicted relationships have, however, provided little evidence that memory differences underlie differences in overprecision. While Hansson *et al* (2008) found support for STM predicting interval width in one task, work by Bruza *et al* (2008) and Kaesler *et al* (2016) failed to find any such effect and the latter, instead, found evidence for Yaniv and Foster's (1995) alternative informativeness–accuracy trade-off explanation.

In short, while consideration of the relationships between memory and biases seems to have significant potential, this has yet to be realised—or extensively tested.

10.2.2 Smart decisions

Just like with expertise, it seems natural to people that smarter people will tend to be less susceptible to biases. This belief makes sense in terms of the two systems approach to heuristics and biases adopted by many researchers in the field (described in chapter 2). The idea here, as you may recall is that, when presented with a problem, a person's intuitive system 1 processes will attempt to provide a solution, and only when it cannot or an error is detected will their rational system 2 processes kick in. Both the detection of an error—and the quality of responses generated by system 2 once activated—seem to reflect abilities that people would think of as intelligence.

The general position within the heuristics and biases literature, however, is that 'intelligence' is only weakly related to bias susceptibility. Frederick's (2005) paper, for instance, finds no correlations between intelligence and the various biases he examined higher than 0.20. Stanovich and West (1998), likewise found only weak correlations between cognitive ability and overconfidence (0.20), hindsight bias (0.24) and syllogistic reasoning (0.34); and, in later work, argued that the true correlations were even lower, as previous experimental methods often relied on within-subjects designs (where one participant sees a number of different tasks and could, potentially, deduce the underlying rationale of a set of questions), leading them to conclude that intelligence and decision-making ability are, largely, independent (Stanovich and West 2008).

Before accepting this conclusion, however, there are two very important questions to consider. First, what exactly do we mean by 'intelligence'? Second, how was this measured in these studies?

Intelligence is a very broad term that can includes a person's reasoning ability, pre-existing knowledge, and their ability to recognise and solve a variety of problems. Since the foundation of psychology as a scientific discipline, however, this very breadth has resulted in different people being able to, meaningfully, use very different definitions for what counts as 'intelligence'.

For instance, since the early twentieth century, some researchers have tended to regard intelligence as a single trait—general intelligence or simply 'g'—underlying performance across a wide range of fields (Spearman 1904) whereas others have argued that intelligence is better thought of as a set of largely independent aptitudes or abilities (e.g. Thurstone 1924).

Work in psychometrics (the measurement of human mental traits) since the 1990s, however, has increasingly pointed towards a hierarchical structure. The most widely accepted model of intelligence is, thus, now the Cattell–Horn–Carroll (CHC) model (McGrew 2009), which posits—supported by Carroll's (1993) extensive meta-analyses—a general intelligence trait that correlates with a set of specific intelligences as shown in figure 10.1.

Looking at the figure, one sees that two of the listed 'broad abilities' are aspects of memory, but intelligence comprises another eight factors reflecting different aspects of people's cognitive abilities—from reasoning to reaction speed—all of which are positively correlated, reflecting a general intelligence factor. That is, people with better memories tend to have faster reaction times, higher processing speed, better comprehension, etc.

With this as background, we can return to the consideration of the role of intelligence in bias susceptibility and perhaps the first thing to leap out from the literature is the fact that amongst the relatively little work on biases that has looked at individual differences in intelligence, almost none considers the implications of the accepted CHC model.

That is, while there are exceptions to the aforementioned tendency for bias research to ignore individual differences, even these tend to use 'blunt' measures of

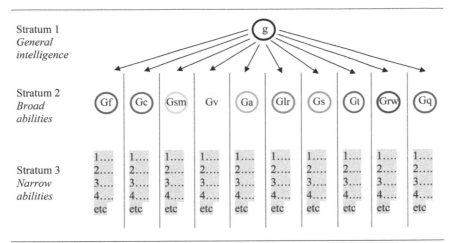

Stratum 1 / General intelligence

Stratum 2 / Broad abilities

Stratum 3 / Narrow abilities

Note: Broad abilities correlate with g at between ~0.6 and 0.9. Gf = fluid intelligence; Gc = comprehension/knowledge; Gsm = short term memory; Gv = visual processing; Ga = auditory processing; Glr = long term storage and retrieval; Gs = processing speed; Gt = decision and reaction speed; Grw = reading and writing; and Gq = quantitative ability. The more than 80 identified narrow abilities that correlate with broad abilities are excluded for reasons of space.

Figure 10.1. CHC model of intelligence (after McGrew 2009).

intelligence. For example, the work of Stanovich and West (1998, 2008), while at the forefront of the consideration of individual differences in decision-making biases, has primarily relied on self-reported SAT (college entry) scores as their measure of 'intelligence'. (SAT scores are, in fact, a measure of intelligence rather than scholastic achievement—correlating with tests known to measure Gc and Gf at between 0.72 and 0.82—and self-reported SATs correlate with actual SATs at above 0.8. (Frey and Detterman 2004)).

Even when a direct measure of intelligence has been used—as, for example, by Frederick (2005)—it is typically a measure of general intelligence without consideration of how particular biases are likely to be related to specific, broad abilities. For example, a number of the questions typically used to test for biases are numerically based, as these allow researchers to calculate objectively correct answers. This however, means that, if some form of intelligence is going to help people avoid these biases, Gq (quantitative ability) might be the most likely candidate. If, rather than a direct measure of Gq, however, we use an indirect (self-reported) measure of something (SAT) that correlates with Gc, which correlates with g, which, finally correlates with Gq—should we be surprised that our measure is not as predictive as we might hope?

Thus, the conclusion for the relationship between intelligence and susceptibility to decision-making biases is, in some way, simply an expansion of the comment on memory in section 10.2.1—current work offers little support for the idea that more intelligent people are less susceptible to biases in decision making, but this may reflect insufficient work in this area rather than a conclusion based on exhaustive examination.

10.2.3 Decisions with personality

Other than intelligence, perhaps the central focus of psychometric work is the consideration of personality and, as was the case for intelligence, various competing theories about the structure of personality have been winnowed down to a single front-running theory that best matches the evidence from decades of research gathered across a variety of cultures world-wide. In this case, it is the 'big 5' personality variables (Costa and McCrae 1997)—commonly remembered using the acronym OCEAN: openness, conscientiousness, extraversion, agreeableness and neuroticism. Each also includes a variety of sub-scales or 'facets'—as shown in table 10.1—that describe specific tendencies contributing to these variables. For instance, extraversion includes as facets both sociability and assertiveness—somewhat distinct attributes that both contribute to our overall perception of someone as extraverted.

While personality has possibly attracted less attention than intelligence within the decision-making literature, looking at the factors and facets in the table, one can see that there are several aspects of personality that seem, potentially, at least, to relate to how people make decisions. For instance, conscientiousness, with its facets of dutifulness, achievement, striving and deliberation, suggests a person who scores highly on this trait is likely to be thorough and vigilant. Thus, they might be more likely to check their responses and, as a result, be more likely to notice and correct their own errors. That is, high conscientiousness may predict more switching from system 1 to system 2, which can protect against biases in some situations.

Conversely, however, high conscientiousness people have also been shown (Musch 2003) to be *more* susceptible to hindsight bias (see chapter 8). The suggestion here is that these people more conscientiously update their memories with new information—spending more time trying to construct causal explanations of observed events—which results in them being more likely to believe that the event was easily predictable and, thus, that they did so.

Table 10.1. Big 5 personality factors and facets (after Costa and McCrae 1992).

Factor	Facets		
Openness	Fantasy	Feelings	Ideas
	Aesthetics	Actions	Values
Conscientiousness	Competence	Order	Dutifulness
	Achievement striving	Self-discipline	Deliberation
Extraversion	Warmth	Activity	Assertiveness
	Gregariousness	Excitement seeking	Positive emotions
Agreeableness	Trust	Altruism	Modesty
	Straightforwardness	Compliance	Tender-mindedness
Neuroticism	Anxiety	Depression	Impulsiveness
	Hostility	Self-consciousness	Vulnerability

Other potential relationships between personality and biases are suggested if one digs a little deeper into the meaning of the facets. For example, the openness ideas facet reflects how open to new ideas a person is. That is, their intellectual curiosity and how willing they are to change their views in light of new information. This could be linked to a variety of biases relating to how people update their opinions— for example, anchoring and hindsight biases. McElroy and Dowd (2007) tested the first of these predictions and found evidence that people high in openness were more susceptible to anchoring.

Thus far, however, the majority of this work has been done at the factor rather than the potentially more informative facet level, and some has been done with alternative personality tests like the Myer–Briggs, which (despite its popularity) psychologists generally regard as problematic for a number of reasons (for an example of such a test and accompanying criticism, see Welsh *et al* 2011).

In general, however, the literature linking personality and decision making makes an important point (that holds true for intelligence and decision styles as well): because different biases result from different cognitive processes, there is unlikely to be a single personality 'type' that protects against 'bias' in general. Rather, individual personality traits or facets may predict greater or lesser susceptibility to particular biases and so, if we are interested in who will make the best decisions, we need to consider which decision we are interested in and identify the biases that we are most worried about affecting that decisions. Only then could we start to think about selecting people based on their traits.

10.2.4 Stylish decisions

The final area of individual differences relating to decision-making ability and bias susceptibility is 'decision styles'. These differ from the above areas of intelligence and personality in that they are a more recent idea (several decades old rather than a century or more) and, as a result, are yet to converge on a single underlying theory in the way that intelligence and personality now seem to have. Instead, this field consists of a somewhat disparate set of measures that seek to distinguish between how people *prefer* to make decisions.

As a result of this focus, decision styles have ended up straddling the region between personality (measures of people's preferred behaviours) and intelligence (which includes the *abilities* required to make decisions). This means that a number of these measures correlate with both intelligence and personality measures—yet they still retain some predictive power after these have been accounted for.

10.2.4.1 Reflect on this
Before going on, please read the questions below and write down your answers to each.

A. If you're running a race and you pass the person in second place, what place are you in?
B. A farmer had 15 sheep and all but 8 died. How many are left?

C. Emily's father has three daughters. The first two are named April and May. What is the third daughter's name?

10.2.4.2 Cognitive reflection

The above are questions from Thompson and Oppenheimer's (2016) expansion of Frederick's (2005) cognitive reflection test (CRT) decision style measure. These sorts of questions are argued to distinguish between people's level of 'cognitive reflection'—that is, to measure whether a person prefers to go with their intuitive responses or to reflect on their initial thoughts. As each question has an immediately intuitive response that is, on reflection, easily shown to be wrong, getting the answer right therefore indicates a greater propensity to reflect on one's responses. (For each of the questions, the intuitive (I) and correct (C) answers are: A. 1st (I) and 2nd (C); B. 7 (I) and 8 (C); and C. June (I) and Emily (C).)

The original CRT has been shown to relate more strongly than intelligence to a variety of biases (Frederick 2005, Toplak *et al* 2011), prompting a great deal of attention. This attention has, however, pointed out three core problems with the CRT. The first is that the measure consists of only three questions; the second, that its very popularity has resulted in its questions becoming extremely well known (see, e.g. Welsh and Begg 2017); and, finally, all three of the original questions were numerical with the result that it has been questioned whether the CRT is actually measuring a decision style or just quantitative ability (intelligence) (see, e.g. Weller *et al* 2013, Welsh *et al* 2013).

This has resulted in updated questions like those above and such work has resulted in a growing consensus that CRT-style questions do measure a 'cognitive reflection' trait in addition to any input from intelligence or personality, and that people higher in this trait are less likely to display biases that result from a tendency to rely, uncritically, on intuitions.

10.2.4.3 Rational and intuitive

While, outside of psychology, the CRT is perhaps the best-known decision style measure, within the field there are other measures with longer pedigrees and equally impressive connections to decision-making abilities.

'Need for cognition' (Cacioppo and Petty 1982), which measures people's preference for engaging in reasoning, is perhaps the best studied decision styles measure and has been linked with differential susceptibility to a wide range of biases (see, e.g. Petty *et al* 2009). Specifically, people with a high need for cognition spend more time engaged in metacognition (thinking about their own thoughts) and, thus, are less susceptible to biases resulting from reliance on system 1 (intuitive) thinking, show less stereotyping and less of a halo effect. Conversely, however, they are *more* susceptible to biases that result from system 2 processes—such as hindsight bias and false memory creation (chapter 8) or even, potentially, confirmation bias.

This observation may sound familiar—echoing the conclusions about conscientiousness above. This is unsurprising as NFC has been shown to correlate with both intelligence and conscientiousness—while still retaining its own predictive power. In effect, while people who are smarter and more conscientious also prefer to engage in reasoning, there is some, additional variance being explained by the NFC questions that improves its predictive power—a common observation from decision styles measures that focus on the difference between preferences for rational and intuitive decision making (see, e.g. Hamilton *et al* 2016). More recent work on rational/ intuitive decision styles have also, however, pointed out an important distinction not drawn in the original NFC—that is, that preferences for rational and intuitive reasoning are not the end-points of a single trait but rather two, separable traits— preference for engaging in rational thinking and preference for intuitive thinking— that are only weakly correlated. That is, a person can enjoy and have trust in *both* their rational and intuitive thinking.

10.3 Implications for scientists

The three strands of individual differences work have implications for thinking about the biases that are likely to affect scientists. Specifically, if we think about the traits required for someone to become a scientist, this may give us insight into the types of biases that scientists may be most susceptible to. Drawing on our stereo- types of scientists, for example, we would assume that they are likely to be smarter than average and to be higher in conscientiousness and need for cognition (NFC)— as both attention to detail and logical reasoning are important skills in science. Thus, we would expect scientists to be somewhat less affected by biases resulting from overreliance on system 1, intuitive, processes and less affected by stereotypes and halo effects. Conversely, effects like hindsight bias and confirmation bias could, actually, be expected at higher than typical levels as these can be magnified by a person's ability to reason. In contrast, when communicating with non-scientists, a scientist should remember that their audience's individual traits and tendencies may result in a greater likelihood of particular biases. One could, as a result, identify the most likely biases and take steps to address these as discussed elsewhere in this book.

Of course, it is important to remember that the above are general tendencies or trends, rather than hard rules and, even in cases where bias susceptibility is lessened, it may still exist at significant levels (as will be discussed further in chapter 11). Perhaps more important is the fact that, while these population-level correlations may predict differences between scientists and non-scientists, they are unlikely to be very helpful within the scientific community as a means of, for example, selecting better decision makers. This is because range truncation (recall figure 9.1 and accompanying discussion) tends to reduce the strength of relationships in already selected segments of a population and to become a scientist, a person has had to undergo rigorous selection in terms of intelligence, etc.

As an example of this, the average CRT score in Frederick's (2005) original study was 1.24 (out of 3). A recent sample of engineers and geoscientists unfamiliar with the CRT questions, by comparison, averaged 2.12—with a median score of 3 (Welsh and Begg 2017). (Those who recalled having seen *any* of the CRT questions previously averaged 2.65.) At these levels, the CRT retained little predictive power for any of the seven biases tested (anchoring, overconfidence, framing, sample size invariance, conjunctive/disjunctive events bias, selection bias and illusory correlations).

10.4 Conclusions

People differ in a variety of measurable ways and differences in their abilities and preferences predict some of the differences that we observe between people in terms of their susceptibility to bias. Knowing which traits affect which biases, however, requires us to think about the cognitive processing that underlies decision making and the ways in which this can lead to bias. With this information in mind, we can make reasonable predictions about whether a trait is likely to predict better or worse performance on tasks of this sort—that is, more or less susceptibility to that, specific, bias.

This approach, however, needs to be taken with a grain of salt. Not because the core idea is wrong but, rather, because this conjunction between the cognitive and psychometric fields of psychology has received less attention than it is, perhaps, due and, as a result, its findings are sparser and less reliable than they could be.

It should also be kept in mind that the observed relationships tend to be relatively weak (a function, perhaps, of the above point) and, so, the existence of a relationship between a trait and a bias does not imply that people high in that trait will never show the bias. Rather, it is possible that they still, consistently show it—just less often or more weakly than other people.

Finally, perhaps the most important take-away from this work is that there is no single trait that protects against all biases. Because biases result from a wide variety of different cognitive processes, a trait that protects against one type of bias can open a person up to other biases. In short, while we may all be different, we are all likely to be biased in some way.

References

Bruza B, Welsh M and Navarro D 2008 Does memory mediate susceptibility to cognitive biases? Implications of the decision-by-sampling theory *Proc. 30th Ann. Conf. Cognitive Science Society* ed B C Love, K McRae and V M Sloutsky (Austin, TX: Cognitive Science Society) pp 1498–1503

Cacioppo J T and Petty R E 1982 The need for cognition *J. Person. Social Psychol.* **42** 116–31

Carroll J B 1993 *Human Cognitive Abilities: A Survey of Factor-Analytic Studies* (Cambridge: Cambridge University Press)

Costa P T and McCrae R R 1992 *NEO PI-R Professional Manual* (Odessa, FL: Psychological Assessment Resources)

Deary I J 2001 Human intelligence differences: towards a combined experimental–differential approach *Trends Cogn. Sci.* **5** 164–70

Englich B and Mussweiler T 2001 Sentencing under uncertainty: anchoring effects in the courtroom *J. Appl. Social Psychol.* **31** 1535–51

Englich B, Mussweiler T and Strack F 2005 The last word in court—a hidden disadvantage for the defense *Law Hum. Behav.* **29** 705

Englich B, Mussweiler T and Strack F 2006 Playing dice with criminal sentences: the influence of irrelevant anchors on experts' judicial decision making *Person. Social Psychol. Bull.* **32** 188–200

Frederick S 2005 Cognitive reflection and decision making *J. Econ. Persp.* **19** 25–42

Frey M C and Detterman D K 2004 Scholastic assessment or g? The relationship between the scholastic assessment test and general cognitive ability *Psychol. Sci.* **15** 373–8

Hamilton K, Shih S I and Mohammed S 2016 The development and validation of the rational and intuitive decision styles scale *J. Person. Assessm.* **98** 523–35

Howard G 1983 *Frames of Mind: The Theory of Multiple Intelligences* (New York: Basics)

Juslin P, Winman A and Hansson P 2007 The naive intuitive statistician: a naive sampling model of intuitive confidence intervals *Psychol. Rev.* **114** 678

Kaesler M P, Welsh M B and Semmler C 2016 Predicting overprecision in range estimation *Proc. 38th Ann. Conf. Cognitive Science Society* ed A Papafragou *et al* (Austin, TX: Cognitive Science Society) pp 502–7

Kahneman D and Klein G 2009 Conditions for intuitive expertise: a failure to disagree *Am. Psychol.* **64** 515

Klein G A and Zsambok C E (ed) 1997 *Naturalistic Decision Making* (Leipzig: Laurence Erlbaum)

Malhotra V, Lee M D and Khurana A 2007 Domain experts influence decision quality: towards a robust method for their identification *J. Petrol. Sci. Eng.* **57** 181–94

McCrae R R and Costa P T Jr 1997 Personality trait structure as a human universal *Am. Psychol.* **52** 509

McElroy T and Dowd K 2007 Susceptibility to anchoring effects: how openness-to-experience influences responses to anchoring cues *Judgm. Decis. Making* **2** 48

McGrew K S 2009 CHC theory and the human cognitive abilities project: standing on the shoulders of the giants of psychometric intelligence research *Intelligence* **37** 1–10

McKenzie C R, Liersch M J and Yaniv I 2008 Overconfidence in interval estimates: what does expertise buy you? *Organ. Behav. Hum. Decis. Process.* **107** 179–91

Miller G A 1956 The magical number seven, plus or minus two: some limits on our capacity for processing information *Psychol. Rev.* **63** 81

Murphy A H and Winkler R L 1984 Probability forecasting in meteorology *J. Am. Statist. Assoc.* **79** 489–500

Musch J 2003 Personality differences in hindsight bias *Memory* **11** 473–89

Petty R E, Briñol P, Loersch C and McCaslin M J 2009 The need for cognition *Handbook of Individual Differences in Social Behavior* (New York: Guilford) pp 318–29

Spearman C 1904 "General intelligence", objectively determined and measured *Am. J. Psychol.* **15** 201–92

Stanovich K E and West R F 1998 Individual differences in rational thought *J. Exp. Psychol.* **127** 161

Stanovich K E and West R F 2008 On the relative independence of thinking biases and cognitive ability *J. Person. Social Psychol.* **94** 672

Stewart N, Chater N and Brown G D 2006 Decision by sampling *Cogn. Psychol.* **53** 1–26

Thomson K S and Oppenheimer D M 2016 Investigating an alternate form of the cognitive reflection test *Judgm. Decis. Making* **11** 99

Thurstone L L 1924 *The Nature of Intelligence* (Abingdon: Routledge)

Toplak M E, West R F and Stanovich K E 2011 The cognitive reflection test as a predictor of performance on heuristics-and-biases tasks *Memory Cogn.* **39** 1275–89

Webster D M and Kruglanski A W 1994 Individual differences in need for cognitive closure *J. Person. Social Psychol.* **67** 1049

Weller J A, Dieckmann N F, Tusler M, Mertz C K, Burns W J and Peters E 2013 Development and testing of an abbreviated numeracy scale: a Rasch analysis approach *J. Behav. Decis. Mak.* **26** 198–212

Welsh M B, Delfabbro P H, Burns N R and Begg S H 2014 Individual differences in anchoring: traits and experience *Learn. Indiv Differences* **29** 131–40

Welsh M B, Alhakim A, Ball F, Dunstan J and Begg S 2011 Do personality traits affect decision-making ability: can MBTI type predict biases? *APPEA J.* **51** 359–68

Welsh M and Begg S 2017 The cognitive reflection test: familiarity and predictive power in professionals ed G Gunzelmann *et al Proc. 39th Annual Cognitive Science Society Meeting* (Austin, TX: Cognitive Science Society)

Welsh M, Burns N and Delfabbro P 2013 The cognitive reflection test: how much more than numerical ability? *Proc. 35th Ann. Conf. Cognitive Science Society* ed M Knauff *et al* (Austin, TX: Cognitive Science Society) pp 1587–92

Welsh M B, Rees N, Ringwood H and Begg S H 2010 The planning fallacy in oil and gas decision-making *APPEA J.* **50** 389–402

Yaniv I and Foster D P 1995 Graininess of judgment under uncertainty: an accuracy-informativeness trade-off *J. Exp. Psychol.* **124** 424

Part III

Implications and solutions

IOP Publishing

Bias in Science and Communication
A field guide
Matthew Welsh

Chapter 11

Warp and weft: publication bias example to weave it all together

The previous seven chapters have each focused on a bias or set of biases that seem to arise due to similar cognitive processing—or, at least, are thematically linked—attempting to explain the general principles by which one might observe or be affected by particular biases. This chapter, by comparison, takes more of a case study approach—looking at areas of particular concern to science and drawing threads from various of the biases together in order to explain how we end up in situations where bias has a marked impact on people's behaviours and beliefs and, as a result, on the pursuit of science.

Specifically, examples have been chosen to highlight the main concerns of this book: how biases in decision making can affect the conduct of science; how they can affect the communication of scientific findings; and the implications of this for how we should behave and communicate.

11.1 Biased science

While the goal of science is uncovering the true nature of the world, scientists are people and, as described previously, suffer from the same limited cognitive abilities as other people. Given this, we must expect scientists and, as a result, science to suffer from biases resulting from these limitations. By way of example, herein, we will examine two central concerns about the conduct and, in particular, the communication of science. As has typically been the case in this book, we are not focussing on deliberate attempts to falsify results or influence which work gets published; rather, the examination will include cases where largely unconscious biases affect the behaviour of otherwise well-meaning individuals.

11.1.1 Publication bias

The phrase 'publication bias' could cover a range of sins but has come to be used, specifically, to refer to researchers' tendencies to decide whether to pursue publication of a finding depending on what the research found and how interesting or important is seems. In particular, the observation that researchers are far more likely to seek the publication of 'positive' results—that is, those that have found new effects in line with their predictions. This is not a newly identified problem, with Rosenthal (1979) discussing psychology's lack of tolerance for 'null' results (results that fail to meet the arbitrary $p < .05$ standard for null hypothesis significance testing) as resulting in what he called the 'file drawer problem'. This refers to the idea that a researcher's filing cabinet could be filled with studies that they have conducted but which failed to find a statistically significant result. Rosenthal, in fact, pointed out the most extreme interpretation of this—where the published literature might consist entirely of the 5% type 1 errors (i.e. random variations being mistaken for effects) expected from the conventional $p < .05$ significance standard meaning that, for each published paper, 19 more might be languishing in drawers.

While no one believes the situation in any area of science is quite that bad, there are a number of human tendencies that push people in this direction and evidence from metanalyses suggests that this does affect the scientific literature in a significant way. Kicinski (2013) for instance, performed a meta-meta-analysis—an analysis of 49 previous meta-analyses of medical research—and found statistical evidence that 42 (86%) of these displayed a bias towards publication of positive results. About a third of the meta-analyses (16/49) showed strong evidence, with positive results being more than twice as likely to be reported—suggesting that a significant number of experimental results do end up in the file drawer.

Fanelli (2010), in a large, cross-disciplinary meta-analysis, found that positive publications constituted 84% of published, scientific works—ranging from a low of ~70% in space science up to a high of above 90% in psychology/psychiatry. Analysis of the statistical power of studies, however, fails to match these higher success rates—that is, given the experimental design and results as described in the literature, we should expect fewer successful experiments than are observed—supporting the idea that unsuccessful results are being suppressed (for a discussion, see Bakker *et al* 2012).

Possible reasons for this are discussed below. Specifically, the role of confirmation bias (chapter 5), hindsight bias (chapter 8) and sample size invariance (chapter 4) are considered in terms of how they affect a scientist's decisions regarded how to frame and disseminate their work. (The important role that editorial decisions play in reinforcing this bias is discussed in the following section 11.1.2.)

11.1.1.1 Constructing causal stories

The scientific process relies on the construction of falsifiable hypotheses. That is, propositions about the world that can be empirically or experimentally tested and discarded if they do not accord with observations. This iterative process of hypothesis generation and rejection should result in a set of hypotheses that are yet to be rejected—our current, best explanations for the events we are observing.

This means that even long-established theories, such as evolution by natural selection, general relativity and so forth, are regarded, within science, not as 'proven true' but rather, 'yet to be falsified'.

This is not, however, exactly how the process occurs in people's minds. Instead, typically, the process will run in the other direction. People might make observations that are at odds with a current theory (or with the absence of a theory in the case of a newly discovered behaviour or event)—which prompts them to wonder whether there is a better, alternative explanation. This generative process, however, requires that the hypothesiser create a causal explanation for these observations. From that stage, the focus tends to be on seeing whether the new theory provides a better explanation than the old—that is, evidence is collected to try to show the new explanation to be true (or, at least, truer than the old).

This is the underlying motivation behind null hypothesis significance testing—where the goal is not to falsify the newly created alternative hypothesis but rather its mirror image 'null hypothesis'. That is, the null hypothesis is defined as the absence or opposite of the effect that the new hypothesis is trying to explain, and the observational or experimental work is focused on trying to gather evidence that the null hypothesis is false and thus, logically, that the alternative hypothesis is true.

Why this occurs relates back to confirmation bias and people's tendencies to believe things more strongly when a good causal story has been developed to explain the effect. As described in chapter 5, people's natural tendency is to seek evidence that conforms with their beliefs and, in generating the alternative hypothesis, a scientist has invested some degree of belief in it. This makes it natural to seek evidence *for* rather than against this new hypothesis—particularly in light of the fact that the researcher clearly has a reason or evidence causing them to doubt any current explanation (otherwise, why would they be conducting the research?). Kerr (1998) cites Mahoney (1977) as providing evidence that scientists are, in fact, susceptible to confirmation bias in hypothesis testing, with 89% believing a confirmatory test of a hypothesis was valid but only 39% recognising that a disconfirmatory one was.

This results in people believing that null results are failures—as it suggests a world where no good causal story exists. That is, the impression that a researcher may be left with is that their negative result, in failing to provide a better explanation, is not adding anything to the literature or that their experiment is, in some way, flawed. A natural reaction to this may be to either file the experiment away as a failure or to 'tweak' their design and then try again.

This approach (running multiple iterations of the same or very similar experiments to 'iron out problems', of course, increases the likelihood of finding a significant result as the chance of a type I error (falsely rejecting the null hypothesis)) is:

$$1-(1-\alpha)^x, \tag{11.1}$$

where α is the selected significance level of the statistical test (e.g. 0.05) and x is the number of tests run (i.e. 0.05 for one test, ~0.10 for two tests, ~0.14 for three tests, and so on).

Either choice results in a preponderance of positive results being sent for review and thus contributes to publication bias.

11.1.1.2 HARK, can you hypothesise that?

Another effect resulting from people's desire to construct causal explanations is called 'HARKing', with HARK being an acronym for 'hypothesising after results are known' (Kerr 1998). That is, rather than writing up a paper that accurately reflects the process through which an experiment was designed and the *a priori* hypotheses tested, this refers to a tendency to examine the results and then go back to restate existing hypotheses so as to better accord with observed results—or to generate post hoc hypotheses and write the paper as if these were the original, guiding intent behind the experiment.

To some extent, this seems natural—to look at the results and generate new hypotheses on the basis of observation is clearly a key role of a scientist. Kerr (1998), in fact, includes evidence that a significant portion of social scientists regard some form of HARKing as appropriate. Some authors (Bern 1991) have suggested that this form of revision is, in fact, beneficial as it clarifies scientific communication. Rather than have to explain all of our failed ideas and false starts, we can simply present the simplified 'final' story, which more clearly communicates what we believe we have discovered or learnt.

However, doing so and then treating the current experiment as if it were designed to test this new hypothesis rather than designing a new experiment to test it, runs the risk of increasing publication bias as described below.

First, as discussed in chapter 9, simplifying a story increases its fluency which, in turn, increases how likely we are to believe it. While this doubtless has benefits in terms of convincing editors of the veracity of our finding (as discussed below) it means that, as we prune away failed hypotheses and simplify our communications, we are likely to start believing in our own simple, core message more strongly. That is, by removing failed hypotheses, we can remove our doubt which, in turn, makes it easier to focus in on the story we are telling and look for confirming evidence (confirmation bias, cf chapter 5) rather than having to think about the complex reality which might contain more, potentially disconfirming, evidence.

Second, the simpler, causal explanations generated by this pruning are easier to project backwards in time—via hindsight bias (chapter 8)—leading to us believing that these post hoc hypotheses were part of our original justification for designing and running the experiment. This is of particular concern when the writing of a research report does not begin until after the experiment has been concluded. A process by which the methodology is accurately recorded and written up in advance of the experiment should, therefore, protect against this (as described in the section on pre-registration below).

Third, this process increases the chance of type 1 errors as the (e.g.) 5% expected error rate is based upon a single, pre-stated hypothesis. If, however, we are allowed to look at the data and create a generate new hypotheses that might better explain the results we see—before selecting the best of these to present as if it were the only hypothesis we considered—then we are multiplying the number of hypotheses (without any clear limit) and the probability of a type I error rises as per equation (11.1) above.

Finally, this leads to bad science as an experiment's design tends not to exactly match the new hypothesis. Kerr (1998), in fact, identifies this a means of recognising HARKing in papers—the absence of control conditions that would be obvious had the presented hypothesis actually been the one being tested.

Overall then, HARKing is a cognitively attractive approach to conducting science but one which increases the likelihood of finding (and, as described below, having published) a positive result that does not, in fact, reflect a real relationship—contributing to publication bias.

11.1.1.3 The significance of power

A third factor contributing to publication bias relates back to a point raised in section 11.1.1. Specifically, the mismatch between observed effect sizes and the statistical power of experimental designs, which is fundamentally linked to sample sizes. In short, statistical analyses indicate that the typical sample sizes in some fields are too small to provide the statistical power observed in the literature as a whole. For example, Bakker et al (2012) suggest that, in psychology, with a moderate effect size (e.g. a Cohen's d of 0.5, which reflects two groups whose means differ by half a standard deviation) and a total sample size of 40 (i.e. 20 per group), statistical power will be around 0.35—reflecting a 35% chance of finding this difference if it does, in fact exist. If the literature consists of markedly more positive results than power analyses suggest (which it does), this is evidence of a file drawer effect.

This dependence on small samples is also problematic in other ways. Specifically, smaller studies tend to produce more extreme results—as described in our discussion of sample size invariance (chapter 4). As a reminder, the law of large numbers means that larger samples are more 'typical' of the population from which they are drawn, while smaller samples are more likely to deviate from the true population parameter values. People, however, often fail to draw this distinction—weighing evidence with only a limited appreciation of the effect of sample size (e.g. adjusting their beliefs as if sample size increases logarithmically rather than linearly as shown in Welsh and Navarro 2012). This means that people are willing to place far more trust in the results of a single, small study than the statistical evidence warrants—particularly if the story a paper tells is clear and cogent.

Of course, the extent to which sample size is a concern will vary markedly by field. In areas where experiments can be run repeatedly and millions of data points collected (sub-atomic physics, perhaps), this may not pose any problems. In contrast, in clinical trials and case studies in medicine and pharmacology, the impact of limited sample size could be at similar or higher levels than described for psychology above, which is likely to combine with the previously mentioned effects to exacerbate publication bias.

11.1.1.4 Test–retest?

All of the above suggests a problem with the conduct of science that is leading to a preponderance of positive results being published in the literature. Science, however, should be self-correcting. If a hypothesis does not, in fact, account for observations of the real world, then, over time, experimental results should accumulate that do not support it and the hypotheses should be abandoned—with earlier, positive results being written off as due to chance.

Without delving too deeply into the philosophy of science, however, it should be noted that this is not always as simple as it sounds for a variety of reasons (see, e.g. Kuhn 1975). Of particular relevance herein, however, is the observation that, in many fields of science, replication is not as highly regarded as 'discovery'. As a result, unless a researcher has a good reason to believe, for example, that a finding is erroneous, they may be better off (career-wise) spending their time searching for new discoveries of their own rather than double-checking the work of others. That is, if a published result seems to provide a reasonable explanation for an effect (where reasonable in the minds of most people includes all of the biases described up until this point), most researchers will not see any benefit in attempting to replicate it (for a discussion see, e.g. Pashler and Harris 2012).

This approach, however, ignores the likelihood of new results being type I errors. Begley and Ellis (2012), for example, describe results from an analysis of published findings from the biotechnology firm, Amgen. Of 53 'landmark' findings only six (11%) were confirmed in replications. However, papers published after the landmark papers were just as likely to cite papers where replication has failed as successfully replicated ones. That is, much of the literature moved on *as if* the effects were real.

Even when replications are conducted, our preference for novelty (the same preference that drives the media landscape's obsession with presenting the unusual to consumers noted in the discussion of biases resulting from the availability heuristic in chapter 8) mean that these attempts tend to be conceptual rather than direct replications. That is, they attempt to replicate the effect under somewhat different conditions or in a different environment or with a different group of people —so as provide a novel twist on the findings. For instance, in Makel *et al*'s (2012) analysis, they found only around 1% of papers could be identified as replications of previous work and, of these, more than 80% were conceptual rather than direct replications.

Any differences between an original result and the findings of a conceptual replication—and, in particular a 'failure' to find the same effect—however, are easily put down to experimenter errors, other differences between samples or misunderstandings of the original papers, leading to them rarely providing a knock-out blow to the previous theory. This gives rise to what Ferguson and Heene (2012) call 'undead theories'—theories that have little empirical support yet which continue to be propagated through the literature and seem unable to be killed by counterevidence.

11.2 Judging peer judgement

The above arguments are all framed in terms of the how biases affect scientists as practitioners. They also, however, have clear implications for the communication of scientific findings as the peer-review process runs on the judgements of scientists. That is, the behaviours of scientists in choosing when and how to present to their work has to be considered in light of how editorial decisions are made by other scientists, as these decisions provide the feedback in terms of how to maximise their chances of being published. Given this, we need to consider the extent to which editorial processes lead to the biases in science described above and elsewhere.

11.2.1 The beautiful bottleneck

Publication is, of course, a competitive process, with multiple papers being considered for a limited number of publication pages in any given journal. This means, of course, that editors and reviewers have to judge which manuscripts contain the most interesting and important findings.

The problem resulting from this is that how we make these subjective judgements is affected by the effects described above. For instance, Giner-Sorolla (2012) argues that the 'aesthetics' of a theory is a major determinant of its success. For a scientific paper, aesthetics refers to both how well written the paper is, how original it is (or how surprising), and how neatly the hypotheses explain the results—that is, how good the causal explanation seems.

The grammatical clarity of course, relates back to the concept of fluency raised in chapter 9— where the likelihood of someone believing something to be true depends on how easily processed it is. Given this, well-written papers are more likely to be believed—on top of the strength of their findings.

The originality or surprise factor of a paper is, similarly, linked to the desire for novelty noted in chapter 8 (and in section 11.1.1.4, above). Novel or surprising occurrences attract and hold our attention more easily and, as a result, a paper that describes an effect that people would not have predicted in advance is more likely to be judged valuable—which can cause problems if the surprising result is, for example, the result of a small sample as described in section 11.1.1.3. This is because such a result is less likely to reflect a real effect and yet is equally compelling to most people. (While papers may be rejected for having insufficient power to have found an effect, they are rarely rejected for having too small a sample when an effect has been found.)

Finally, as noted above, stronger causal stories can often result from post hoc theorising, which removes any messy 'thinking' and failures from the process. Kerr (1998), for instance, found 96% of social scientists agreed that a paper written so as to make hypotheses consistent with findings would have a greater chance of publication than one that presented the original hypothesis, contradictory findings and then included an explicitly post hoc explanation of why this might have occurred. More than half of these scientists (who, it must be remembered, compose the same population as the editors and reviewers of such papers) also indicated that they had received direct advice in a review to do exactly that in order to improve their manuscript. That is, their general belief and actual experience is that HARKing improves publication rates—because HARKing produces cleaner, easier to follow manuscripts.

It also seems likely that the presentation of a clean, successful study would prompt a perception of greater competence than would a description of the true evolution of a paper—with missteps, dead-ends and errors. This is likely to produce a halo effect (cf chapter 10), which will prompt the reader to regard the author(s) in a more positive light generally and increase the likelihood of a positive review.

11.2.2 Who you know

Additional concerns about the potential for halo effects and stereotypes relate to the effects of institutional and author prestige on editorial decisions. While blind review processes are used by many journals, which limit the opportunity for these, the editors themselves are not typically blind to the author's identity and, thus, these biases can affect decisions as to whether the paper is sent out to review in the first place or simply rejected and then the final decision as to whether the paper should be published. (It should also be noted that even the removal of author and affiliation information from a manuscript may not be sufficient to effectively hide an author's identity from an experienced reviewer.)

The question, then, is how much does knowledge of an author's identity actually affect the probability of a paper being accepted? The answer is, of course, complicated due to the methodological difficulties in studying this and confounding factors. As noted in chapter 9, the work of Peters and Ceci (1982) found that only one of the nine previous papers they resubmitted (which were not recognised as resubmissions) was accepted when they altered the authors and institutions to unrecognisable ones. The extent to which this can be taken as typical, however, is arguable given the very small sample size and the typically low acceptance rate at prestigious journals in psychology (significantly lower than in physical sciences, for instance).

Other studies, however, have provided additional evidence along these lines. For example, Garfunkel *et al*'s (1994) study of papers submitted by authors at various medical institutions found that there was clear evidence that short papers (case studies, etc.) submitted by authors at the most prestigious institutions were recommended by reviewers and accepted by editors at significantly higher rates than those from less prestigious institutions (32% compared to 11% for the least prestigious). This study, however, made no attempt to assess the quality of the papers—leaving open the possibility that this could be explained by differences in the quality of the submissions. That said, the study found no evidence of longer papers from more prestigious institutions being more likely to be accepted, which undermines the idea that the authors at these institutions necessarily produce better manuscripts and, instead, requires a somewhat counterintuitive idea that more effort is put into the short papers than the long. A simpler, alternative interpretation, however, could be that a longer paper gives more opportunity for the quality of the paper to overcome any institutional prestige bias—eroding any advantage enjoyed by the high-prestige authors on shorter papers.

Given the above, a safe conclusion seems to be that the prestige of an author or institution *could* act to bias editors and reviewers but that significant work remains to be done before this can be definitively shown—and, even if it is, the peculiarities of different fields may make this differentially problematic. What would actually be required is a large, cross-disciplinary, experimental study in which papers of pre-assessed quality were attributed to famous or unknown authors and institutions prior to being sent for reviews to a large number of reviewers reviewing for journals using both open and blind review processes.

11.2.3 Him, her and them

The same arguments made about author prestige above also apply to gender differences in scientific publication. To the extent that stereotypes affect people's judgements, these have the potential to increase or decrease an editor or reviewer's affinity with an author and, thus, any judgements about the quality of their work. While, again, the complexity of the question has prevented definitive experimental evidence for the effect of this on editorial decisions, there is strong correlational evidence in some areas and related experimental work that suggests it is likely.

Budden *et al* (2008) for example, examined the proportion of female-first-authored papers in the journal *Behavioral Ecology*—prior to and after it introduced a double-blind reviewing policy in 2001. After this change, the total percentage of papers first authored by women rose from 28% to 37%. While correlational rather than causal, the only other obvious explanation would be a change in the demographics of the field, but Budden *et al* demonstrated no such change in the female-first-author rate at similar journals across the same period—leaving the change to double-blind review and the removal of unconscious gender bias as the most likely explanation.

An experimental study that offers additional evidence of the existence of such biases in scientists' decisions was run by Milkman *et al* (2015), who sent emails to more than 6000 academics, posing as students of differing genders and ethnic backgrounds and asking to arrange a meeting to discuss potential supervision. All of the contact emails were identical—except only for the name, which was chosen as a clear marker of ethnicity and gender—and the measure of bias was any difference in the rate of responding to emails. Across ten fields for study, eight showed a significant bias. For example, in health sciences, 67% of 'white male' queries received a response, whereas only 57% of female or minority queries did—a 10% bias in response rate. An effect of the same magnitude was seen in engineering and computing (10%), and smaller but still significant effects in life sciences (7%), physical sciences (5%) and social sciences (2%). Again, this does not require conscious discrimination on the part of the academics— just an unconscious bias towards assisting or associating with people more like them (white and male being the modal description of US academics) resulting in their being less inclined to spend the time responding to queries.

11.2.4 In agreement

A final area of concern for potential editorial and reviewer bias is confirmation bias—the tendency for people to seek out and more easily agree with evidence that supports their own position. Mahoney (1977) conducted an experiment examining the effect of this on reviewers' ratings of the quality of papers with identical methodologies but positive, mixed or negative results, and found that reviewers judged papers that had findings in accordance with the reviewer's own theoretical beliefs more positively. That is, despite the same methodology being described, results that confirmed the reviewers' opinions were favoured.

11.3 Debiasing science

Given the above, it seems fair to conclude that science is being affected by a variety of biases discussed in previous chapters. However, it also seems clear that there are a variety of systematic changes that could limit the effect of these biases and thus improve the efficacy of scientific endeavours. Two of these are expanded on below.

11.3.1 Revising reviews

As noted above (Budden *et al* 2008), blind review has the potential to reduce ethnic or gender bias in the evaluation of manuscripts—and a wide variety of outlets do use this. Additional changes to the review process could also be considered, however, to help counter other biases. For example, when reviewing a manuscript, one could assess a paper's contribution to science based on its introduction and methodology section—prior to going on read the results and discussion. These sections— providing the background and the experimental design—should be sufficient to determine whether the scientific reasoning and approach taken is sound (i.e. whether the experiment is a valuable contribution to science) while avoiding being affected by confirmation bias (if the results agree with the reviewer's own opinion) or outcome bias (where a positive outcome is attributed to good experimental design).

11.3.2 Rewarding replicators

A related approach is one being undertaken by the Center for Open Science. In response to the initial findings of their Psychology Reproducibility Project (Open Science Collaboration 2015), which conducted 100 direct replications of papers from major psychology journals and found that fewer than half of these supported the initial findings. This work has now been extended to other areas of science (e.g. cancer research) and now includes a variety of platforms designed to enable scientists to share their data and pre-prints of papers—so as to provide a better record of how science is actually being conducted than can be gained from looking solely at final publications.

This approach also includes journals in psychology, medicine, neuroscience, biology and ecology (thus far) allowing the pre-registration of projects (for details see, e.g. https://cos.io/rr/#RR). That is, prior to conducting a study, a scientist can provide their experimental plan to a journal for review and, if the project is adjudged valuable, the journal accepts the paper based on this review—irrespective of how the results turn out. This process achieves the same result as the above-mentioned idea of reviewing a paper based only on its introduction and method but has the additional protection of preventing HARKing and other post hoc redesigns. That is, it restores the proper process for science—of keeping post hoc hypothesising as discussion points and directions for future research—as shown in table 11.1.

Table 11.1. Ideal versus biased conduct of science given a negative result.

Ideal	Biased	Bias
Observation	Observation	
Hypothesis	Hypothesis	
Experimental design	Experimental design	
Conduct	Conduct	
Analyse data	Analyse data	
Conclude H was wrong	Conclude experiment was flawed	Outcome bias
Post hoc reasoning	Post hoc reasoning	Hindsight bias
New hypothesis	Revise method with new hypothesis	HARKing
New experimental design	Re-analyse data	Confirmation bias
etc	Report as if this was the original design	p-inflation

Finally, such projects also provide an impetus for conducting impactful, direct replications, which is necessary to avoid the bias against replications noted above. By making such replications part of a larger project, their impact on the literature and thus their value to the individual scientists conducting them is increased.

11.4 Conclusions

The research and arguments presented above may paint a somewhat worrisome picture for those who hold to the ideal of science as an objective pursuit. As noted throughout this book, however, while the ideal of science is rational and clear, it is and always has been conducted by people who share many typically human cognitive limitations and who are situated in complex, social networks that necessitate the use of simplifications like stereotypes for us to navigate them. Thus, it would probably be more surprising if there were no evidence of bias in science and its communication.

Given the preponderance of evidence above, then, it seems a more valuable use of time to assume the existence of biases and act so as to limit the effect that they have on the conduct and communication of science. This can include large-scale, systematic attempts lie the Open Science framework described above and processes that can be undertaken at the individual level in order to safeguard one's own behaviours and communication against biases—which will be the major focus of chapter 12.

References

Bakker M, van Dijk A and Wicherts J M 2012 The rules of the game called psychological science *Persp. Psychol. Sci.* **7** 543–54

Begley C G and Ellis L M 2012 Raise standards for preclinical cancer research *Nature* **483** 531–3

Bern D J 1991 Writing the research report *Research Methods in Social Relations* 6th edn, ed C Judd, E Smith and L Kidder (Fort Worth, TX: Holt) pp 453–76

Budden A E, Tregenza T, Aarssen L W, Koricheva J, Leimu R and Lortie C J 2008 Double-blind review favours increased representation of female authors *Trends Ecol. Evol.* **23** 4–6

Fanelli D 2010 "Positive" results increase down the hierarchy of the sciences *PloS one* **5** e10068

Ferguson C J and Heene M 2012 A vast graveyard of undead theories: publication bias and psychological science's aversion to the null *Persp. Psychol. Sci.* **7** 555–61

Garfunkel J M, Ulshen M H, Hamrick H J and Lawson E E 1994 Effect of institutional prestige on reviewers' recommendations and editorial decisions *JAMA* **272** 137–8

Giner-Sorolla R 2012 Science or art? How aesthetic standards grease the way through the publication bottleneck but undermine science *Persp. Psychol. Sci.* **7** 562–71

Kerr N L 1998 HARKing: hypothesizing after the results are known *Person. Social Psychol. Rev.* **2** 196–217

Kicinski M 2013 Publication bias in recent meta-analyses *PloS one* **8** e81823

Kuhn T S 1975 *The Structure of Scientific Revolutions* 2nd edn (Chicago: University of Chicago Press)

Mahoney M J 1977 Publication prejudices: an experimental study of confirmatory bias in the peer review system *Cogn. Ther. Res.* **1** 161–75

Makel M C, Plucker J A and Hegarty B 2012 Replications in psychology research: how often do they really occur? *Persp. Psychol. Sci.* **7** 537–42

Milkman K L, Akinola M and Chugh D 2015 What happens before? A field experiment exploring how pay and representation differentially shape bias on the pathway into organizations *J. Appl. Psychol.* **100** 1678

Open Science Collaboration 2015 Estimating the reproducibility of psychological science *Science* **349** aac4716

Pashler H and Harris C R 2012 Is the replicability crisis overblown? Three arguments examined *Persp. Psychol. Sci.* **7** 531–6

Peters D P and Ceci S J 1982 Peer-review practices of psychological journals: the fate of published articles, submitted again *Behav. Brain Sci.* **5** 187–255

Rosenthal R 1979 The file drawer problem and tolerance for null results *Psychol. Bull.* **86** 638

Welsh M B and Navarro D J 2012 Seeing is believing: priors, trust, and base rate neglect *Organ. Behav. Hum. Decis. Process* **119** 1–14

IOP Publishing

Bias in Science and Communication
A field guide
Matthew Welsh

Chapter 12

Felicitous elicitation: reducing biases through better elicitation processes

In the preceding chapters (and culminating in the chapter 11 examples), while some attention has been given to methods for reducing particular biases, demonstrations of the ubiquity of decision-making biases have predominated. In this chapter, however, our focus will shift to debiasing strategies—that is, processes or tools designed to limit or eliminate the impact of biases on estimates and decisions.

The discussion of these has been divided into two parts—focussing on how best to debias your own beliefs and decisions and then how best to approach the elicitation of information from other people so as to limit biases in their estimates or responses. (Note, the complementary focus on how to limit or combat bias in other people during communication is the focus of chapter 13.)

12.1 My biases

Throughout this book, questions designed to highlight a number of biases have been discussed. Many of these were also found in chapter 1—the pop quiz—so as to give readers the opportunity to attempt these questions in a naïve state more akin to that in which you would experience biases in the real world. That is, rather than looking at questions designed to highlight overconfidence (e.g.) in the midst of a discussion of confidence and overconfidence in decision making, where such effects might be at the forefront of your mind, having the pop quiz allowed the opportunity to record answers to these questions prior to any specific learning about the types of biases. This has, hopefully, revealed to you the underlying point of this book: that everyone is susceptible to bias under different circumstances and, as a result, everyone who wants to make better decisions needs to acknowledge this and understand how these biases occur.

Here, therefore, we consider the ways in which such biases can be avoided, limited or, at least, recognised in your own thoughts and decisions.

12.1.1 The wise fool

While I hold out hope that some of the material contained in this book will allow you to take active steps to reduce bias in some circumstances, perhaps the primary goal of this book is to raise awareness of how biases affect people's decisions. An obvious question, therefore, is: does knowing about biases actually help? That is, does taking the attitude of the wise fool who recognises how little they know (or how imperfect their knowledge is), result in better decisions?

To some extent, of course, this question is an oversimplification. More precisely, we are really interested in whether knowing about 'biases' helps or whether you need to know about a particular bias; how well you need to understand a bias's mode of action to gain any benefit; under which circumstances such knowledge helps; to what extent it helps; and so on.

Starting at the broadest level, the answer seems to be that having been told about biases in general seems not to result in any improvement in performance on other biases (see, e.g. Welsh *et al* 2006). On the other hand, whether knowledge of specific biases helps or not depends on the characteristics of that bias. Specifically, it seems, the extent to which the cognitive processing that gives rise to the bias is under conscious control and the ease of implementing any counter-measure.

For example, some of the biases arising from our inability to intuitively deal with probability theory (cf chapter 4) are easily countered once a person recognises problems of this type. Consider, for instance, the sample size invariance question (question 3 in chapter 1). Having been reminded about the impact of sample size on the probability of drawing a representative sample from a population, a question in this format is likely to now prompt additional thought and enable you to recognise and avoid the bias. The extent to which this transfers to less obvious situations, however—such as the evaluation of the strength of evidence for alternative hypotheses—will depend on you reminding yourself of people's tendency to ignore sample size differences. That is, you need to remember that additional 'grain of salt' when presented with a causal story supported by only a small sample—regardless of how good the 'story' is.

Likewise, an understanding of how the positive or negative framing (chapter 6) of a question can prompt people to be more risk seeking or risk averse suggests that such risky choices should be reframed to see whether this affects your preference.

In contrast with the above, knowledge of a bias such as overprecision (i.e. overconfidence in range estimation; cf chapter 7) does not allow you to completely overcome the bias but does immediately suggest a path to reducing the effect of the bias—widening one's ranges. Evidence from decades of research (see, e.g. Lichtenstein *et al* 1982, chapter 9 of Morgan and Henrion 1990, Welsh *et al* 2006), however, shows that the benefit of this is limited. That is, it reduces but does not eliminate overprecision. Welsh *et al*'s (2006) results, for example, found people's miscalibration (the difference between the number of ranges they predicted would contain the true value and the number that did) was reduced from 48% to ~21% when people were made aware of their own overconfidence on these sorts of tasks.

Yaniv and Foster's (1995) description of the informativeness–accuracy trade-off, however, explains why this sort of awareness-based debiasing is limited, observing that people's desire to be informative prevents them from seeing any value in ranges as wide as would be necessary to reflect their actual degree of uncertainty. (This is reminiscent of the Dunning–Kruger effect, cf chapter 7, which described the inability of the ignorant to recognise just how little they know. The key here, though, is that, under conditions of high uncertainty, we are all ignorant and thus likely to succumb to this bias and find it hard to believe that our estimates could be so far wrong.)

Finally, some biases seem to be largely unaffected by people's knowledge of them and the poster-bias for this is anchoring. Even when people are made directly aware of the anchoring bias, this seems not to affect their susceptibility to it (see, e.g. Wilson *et al* 1996). This, it seems, is because the cognitive processes prompted by the consideration of an anchor occur sub-consciously—that is, at a level that cannot be consciously affected. Thus, only by using a more complex strategy to prompt additional or alternative cognition can the impact of an anchor be reduced (see, e.g. Mussweiler *et al* 2000, Welsh and Begg 2018).

That said, in the case of all biases, awareness provides an advantage that can be used to good effect. Specifically, awareness allows recognition of instances when your judgement is likely to have been affected by a bias. For example, while knowing about anchoring does not prevent anchors affecting your judgement, it does allow you to recognise situations where your (or other people's) judgements are likely to have been affected by anchors, which enables you to account for its possible effects on your decisions—by recognising the additional uncertainty this brings into estimates, for example. That is, even where awareness of bias does not directly help, it can do so indirectly by ensuring that you are not too confident in the accuracy of your estimates.

12.1.2 Practice makes perfect?

So, if awareness is insufficient to enable avoidance of biases, this suggests that more active steps need to be taken. These could be thought of as 'lifestyle changes' where, rather than just 'knowing' about the effects, you set out to change behaviours in such a way as to limit the role that bias can play on your decisions and judgements.

An obvious first step along this path is practice. That is, deliberately seeking out biases in a controlled environment and learning the extent to which they affect you and methods for reducing them *in situ*. Of course, as discussed earlier (chapter 10), the ability to learn from an environment depends on the characteristics of that environment. This is why meteorologists are well-calibrated: the environment in which they work allows for regular predictions with fast, accurate feedback on the quality of those predictions.

While this is not true for many domains in which scientists have to work, where long delay times between predictions and outcomes are observed, this approach can still bear fruit. This is because the skillset required to, for example, accurately estimate interval widths (a high and low estimate of the length of a project, for example) or judge the likelihood of future outcomes consists of a combination of specialist technical knowledge and the practice in estimation such that one can, for example, intuitively understand what different likelihoods feel like. Given this, it becomes possible to separate these tasks—practicing making these sorts of judgements and decisions.

For instance, Russo and Schoemaker (1992) describe a training program used inside the oil company Shell in response to overconfidence in the assessment of the 'chance of success' (assigned probability of finding oil at a particular location). This is a classic example of an environment ill-suited to learning—with months of years between prediction and outcome in the real world—so, instead, trainee geologists were presented with scenarios based on real but de-identified past data and asked to assign the chance of success. In this way, they could make a series of such assessments and then immediately received feedback on what proportion of the assessed prospects actually contained oil.

Such, simulated environments offer real benefits but, even without this, making and keeping track of your own prediction and how often you are right is likely to be beneficial in improving calibration as it will lead to better understanding of what, for example, an 80% chance 'feels' like. More generally, the role of practice follows from that of awareness. Rather than just being aware of the existence of biases and looking for them when it is clear that they are likely to be having an effect (in the questions of a book about biases for instance), the practice of better decision making is to constantly look for places where bias *could* affect judgements, try to determine what bias it would be, what effect it might have and, then, what you could try to do about his. Perhaps the most important aspects, given the limitations of human memory (chapter 8) is to record your beliefs at the times so that, when data do come in, you have an objective record to compare them to rather than having to try to recollect or reconstruct your beliefs.

12.1.3 A snowflake's chance

Another important means of limiting the effect of bias on your own beliefs and decisions is to adopt what Kahneman and Tversky (1982) dubbed an 'outside' perspective (chapter 7). That is, treating a project, task, etc, as a typical member of some class of similar events and using the distribution of time, cost, success rate—or whatever other metric is desired—to inform the estimates of those metrics for the individual project. This is contrasted to the more typical approach of focussing on the aspects of a project that are unique and special, which suggests that the project can not be compared to others.

In fact, while there may be some projects so ground-breaking and novel that there is no set of comparable projects to use for this process, these are likely to be few and far between and even an imperfect data set may be informative. For example, Tversky and Kahneman's (1982) experience of writing a new text book for a curriculum is, in many ways a unique project: involving a set of academics who had never worked together on such a project before and would probably never do so again; and the development of a new book which would, presumably, contain a unique set of information. One might, therefore, (and the participants did) argue that their project was unique and the experience of other teams was not relevant. The prediction made by the team member who had actually seen a number of text book projects, however, was far more accurate than any of the estimates made by team members who considered only the specific project—suggesting that, despite any unique attributes of the task, its shared attributes were more important.

As an analogy, consider the legendary uniqueness of snowflakes. A person focused on the details of their particular snowflake could be forgiven for expecting that its behaviour or other characteristics might differ markedly from other, distinctly different snowflakes. However, when predicting how one snowflake will react to a change in environment or other effect, the best predictor is probably what the typical snowflake does rather than its surface appearance—so, if a large proportion of similar projects have crashed and burnt, then it is probably fair to judge that yours, too, has a snowflake's chance in hell of succeeding.

12.1.4 Calculate

The third piece of advice of those attempting to reduce the impact of bias on their decisions is to calculate rather than estimate. As pointed out repeatedly in previous chapters, a significant proportion of biases seem to result from the heuristics that people need to use to process information due to their cognitive and memory limitations.

Given this, unless you are confident that your intuitions are well-trained (from lots of experience inside an amenable environment as described in chapter 10) and thus reflect genuine, expert situational awareness, taking the time to write things down and utilise tools that relieve the cognitive load of a problem is likely to improve your decision making.

Decades of evidence starting with Meehl (1954) supports this across a variety of domains including clinical decisions by doctors and psychologists (for a review, see Grove *et al* 2000) and organisational decisions (e.g. hiring, promotion and performance review; for a review, see, Kuncel *et al* 2013). Across the hundreds of studies detailed in these papers, simple calculations or statistical combination of information outperformed expert judgement (or holistic interpretations of data) in more than half of cases, with the majority of the remainder showing a draw and only a small minority reversing this pattern. As described above, this seems to be simply because the process of using a formula prevents cues from being missed or over-or-under weighted by the expert.

For example, a formula near ubiquitously used by hospital staff is the Apgar score (Apgar 1952). This is a simple sum of scores (from 0 to 2) assigned to a baby at 1 and 5 min after birth on five measures, and remembered via the backronym: appearance, pulse, grimace, activity and respiration. This formula replaced previous, more subjective measures of infant health, with low scores predicting a heightened risk of infant mortality and the need for medical intervention. Such has been the success of this simple formula that it is now used world-wide and is credited with improving the swift diagnosis of infant distress.

That is, while obstetricians and midwives have the expert knowledge to diagnose problems, the Apgar allows them to do so swiftly, ensures that they do not miss a cue during the often-stressful birthing environment, and also provides the distributional lens noted above—as the Apgar numbers are widely used and well studied. As a result, if an infant scores an Apgar below 7, a physician knows that starting treatment is necessary—without having to consider all of the innumerable specifics of a case—because there are significant data showing the relationship between Apgar score and infant outcomes.

12.1.4.1 Visible bias

Another area where the use of maths can be extremely useful in combatting bias is in decision making. Decision analysts, for example, recommend the use of tools like multi-criterion decision making (MCDM; for a description of the process see, e.g. Begg and Bratvold 2010) when faced with decisions that have competing objectives and multiple alternatives to select between. This includes almost all hiring decisions, purchases and even decisions like whether to pursue a career in a new location or institute.

Leaving aside some of the complexities regarding how objectives measured on different scales can be compared to one another (e.g. costs in dollars and safety rating on a 1–5 scale) and whether they are equally important, the basic process is relatively simple—requiring objectives to be clearly stated and each alternative scored as to how well it meets each objective. These scores are then summed and the alternative with the highest score should be selected. Looking at table 12.1, one can see, however, why using a spreadsheet is likely to produce better results here than trying to combine all of this information in your head. With three objectives and four alternatives, there are 12 scores that need to be calculated or assigned prior to their summation into four overall scores for the alternatives—a task beyond the ability of human short term memory.

Perhaps the biggest advantage of this approach to decision making, however, lies in its utility for identifying biased or flawed thinking through the transparency of the process. For example, if, after filling in an MCDM template and calculating the best alternative, you realise that that option is, in fact, *not* your preferred option, then this indicates an error has occurred. For instance, you may have left out an objective that is affecting your preference despite not having been included (due to availability, for example) or perhaps that your weighting of the relative values of the

Table 12.1. MCDM template example for academic hiring decision.

Objectives	Alternative 1	Alternative 2	Alternative 3	Alternative 4
1. Publications	$Score_{11}$	$Score_{12}$	$Score_{13}$	$Score_{14}$
2. Teaching	$Score_{21}$	$Score_{22}$	$Score_{23}$	$Score_{24}$
3. Collegiality	$Score_{31}$	$Score_{32}$	$Score_{33}$	$Score_{34}$
Total	$= \Sigma(S_{11}, S_{21}, S_{31})$	$= \Sigma(S_{12}, S_{22}, S_{32})$	$= \Sigma(S_{13}, S_{23}, S_{33})$	$= \Sigma(S_{14}, S_{24}, S_{34})$

objectives does not reflect how you truly value them. This can, thus, allow you to recognise aspects of a decision that you may otherwise have missed, to recognise areas where your assignment of value to alternatives may have been biased, and so on.

Of course, there are also difficulties for decisions of significant complexity, where the number of objective and alternatives is large, or even unknown, requiring sub-decisions about how to create or search for these. However, there seems to be little evidence that attempting to intuit an answer in such a case would lead to better outcomes than a MCDM approach that deliberately omitted some complexity in favour of calculability; and, again, such an approach makes clear which objectives and alternatives have been modelled in a way that less structured decisions cannot.

12.2 Your biases

The above describes methods for recognising or reducing one's own biases and, of course, these approaches are just as valuable for recognising when other people's decisions are likely to be biased. For example, awareness of how and when biases tend to occur will prime you to recognise them when they do appear. An under-standing of the nature of expertise will, likewise, assist you in deciding when and whether to trust another expert's intuitions and whether, as a result, you should trust their estimates or rely, instead, on data relating to similar events to provide your best estimates. Finally, knowing whether an expert's opinion is based on an intuitive combination of evidence or systematic calculation will shed light on the extent to which bias could have affected their estimates, while the use of an MCDM process could even be used to help to reverse engineer their objectives.

There are, however, considerations beyond this when considering how to deal with the possibility of bias in other people—and, specifically, the potential for bias when you are directly seeking someone else's input through a process of elicitation (chapter 3).

12.2.1 Expert elicitors

As noted in chapter 3, one of the primary forms of scientific communication occurs during expert elicitation, when the scientist as an expert is approached in order to bring their expertise to bear on a difficult decision—providing estimates of unknown parameter values, alternatives and so forth. Elicitation thus occurs between scientists in the same field, those in different fields and between scientists and policy makers— wherever another person's expertise is used to reduce our uncertainty.

As discussed throughout this book, however, elicitation carries with it the risk of the same biases we see in any decisions made under uncertainty. Given that expert advice is most needed when uncertainty is highest and the fact that biases generally have their largest effect under conditions of high uncertainty, this area of commu-nication is key if we wish to reduce biases affecting science and scientists.

One method for doing this, is through the use of expert elicitors. That is, rather than relying on the subject matter expert (i.e. the scientist) having the knowledge and skill to avoid bias themselves, ensuring that the person tasked with eliciting the values from them does. That is, someone who has a thorough understanding of the various biases that affect decisions, how and when they occur and how best to ask questions so as to avoid these. This is, of course, a skill-set that many could master but one which almost no one should be expected to possess as the result of expertise in a particular area of science. Thus, it is generally the case that outside training or assistance needs to be sought to fill this gap (and, hopefully, this book will help to set readers on the path, at least).

12.2.1.1 Expert interviews

In various industries, for example, there are a variety of consulting companies who act as expert elicitors—engaging in 'expert interviews' with geoscientists or engineers in order to limit the effect of biases. As an example, Hawkins *et al* (2002) describe an eight-step expert interview process designed to reduce the impact of biases on elicited parameter value ranges relating to oil exploration and recovery. Of these, one (step 7) is specific to particular applications of probability and so will not be included here. The others, however, are generally applicable to any elicitation process designed to limit the effect of bias. For example, the first four steps—which the authors argue should precede the expert examining any analogous data—are to:

1. Counteract motivations bias through discussion of the project, the expert's role in it, and people's tendency to display motivation biases (in effect, making clear that motivation biases will be noticed).
2. Clearly define the variable, thereby ensuring there is no confusion about what the expert is being asked to provide.
3. Draw an influence diagram, to ensure that factors affecting the value of the estimated parameter are discovered and kept in mind.
4. Postulate extreme outcomes, asking the expert to come up with scenarios whereby the parameter values could be shockingly low or unexpectedly high —in an attempt to generate a wide range of possible values rather than allowing the expert to be anchored (see chapter 8) by a 'likely' value or previous data.

Following this and the expert's consideration of any available data, the remaining steps are used to generate the end points of an 80% confidence range and then the expert's best estimate, as follows:

5. Use equivalent bets to establish the 10th and 90th percentiles for the uncertain parameter, specifically by having them select between a game with a one-in-ten chance of winning and betting on the value of the parameter being above or below a possible value for the aforementioned percentiles—as only if the expert has no preference between these games does the estimated value reflect their subjective 10th or 90th percentile.
6. Assess the best estimate by selecting potential values between the 10th and 90th percentiles and asking the expert whether they would prefer to bet on the value being above or below the potential value, with their best estimate being the point at which they become indifferent between the two options.
7. (Not included.)

8. A reality check, where the resulting values are explained to the expert and they are asked to look at these and ensure that they do, in fact, reflect the expert's knowledge.

This is, of course, not the only expert–elicitor-driven process (see, e.g. the Stanford Research Initiative method born out of Spetzler and Stael von Holstein 1975) but shares with others the drawback of being a long, involved process—often requiring 45 min to an hour to elicit estimates for a single parameter. As such, while the general principles can be used, engaging the full process can be too time and effort consuming for day-to-day rather than make-or-break decisions.

12.2.2 Elicitation tools

A way around this is to attempt to design elicitation tools—that is, processes or computer programs built with an understanding of how people remember (chapter 8) and process information (chapters 4, 5 and 7) and, importantly, respond to questions in different formats (as per chapter 6).

For example, Haran *et al*'s (2010) SPIES procedure reduces overprecision (see chapter 7) by forcing the elicitee to assess every possible value of the parameter rather than allowing them to select their own set of possible values. This requires an elicitor to determine the entire range of possible values and divide this into a feasible number of quantiles or bins, but can then be computerised such that the participant is asked to assess the probability of the true value falling into each of these possible regions (and then ensuring that these sum to 1 across all regions). This prevents the expert focussing-in on the area they consider most likely (anchoring, cf chapter 8) and then seeking confirmatory evidence (confirmation bias, cf chapter 5) thereby resulting in wider estimated ranges.

12.2.2.1 *Molehills from mountains*

My own elicitation tool—the MOLE (more-or-less elicitation)—has, similarly, been designed around four key insights drawn from the decision-making literature to reduce overconfidence and limit the effect of anchors (for more detail of its development and underlying processes, see Welsh *et al* 2008, 2009, Welsh and Begg 2018).

First, it recognises people's tendency (described in chapter 8) to swiftly discard regions of possible values that they are not *certain* should be included—so as to avoid being uninformative. Second, it recognises that people both prefer and are better at making relative rather than absolute judgements (see chapter 6). Third, that evidence from the wisdom of crowds literature demonstrates that combining repeated judgements relating to the same parameter produces better estimates (see chapter 3). Then, finally, that any single value can anchor subsequent estimates and thus that this effect needs to be washed out with additional estimates (as noted in section 12.1.1).

The MOLE seeks to accommodate all four insights by, across a number of iterations, randomly selecting paired values from across the range of possible values.

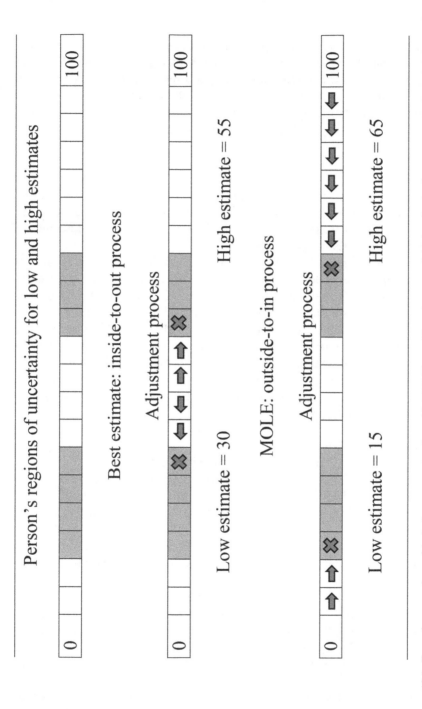

Figure 12.1. Example comparison of typical range construction with MOLE's outside-to-in range elicitation. In the first, a person starts at their best estimate and adjusts outwards until they reach the 'region of uncertainty' (which contains the values they believe are feasible) and then stops. In the MOLE, by comparison, this adjustment process works from the outside of the range in until it reaches the region of uncertainty, retaining more of the person's uncertainty.

This allows the participant to select which of each pair they think is closer to the true value (or more likely) utilising *relative* judgements. It also allows them to *repeatedly* consider the same problem—the most likely location of the parameter value—in such a way as to prevent them from simply repeating their answer, as at no point do they actually state their best estimate but rather have to consider it in light of the two current options. This repeated presentation of options also serves to wash out the effect of any individual *anchor*—forcing the person to consider all of the options presented to them (drawn, you will recall, from across the full range of possible values) in terms of whether they are feasible. Finally, along with each choice between the two options, the elicitee is asked to state how confident they are that the option they have chosen is closer to the true value and only when they are *certain* does the MOLE truncate the range from which it is drawing options. That is, rather than allowing people to only include values they are certain of, the MOLE only eliminates values they are certain are not feasible, as shown in figure 12.1.

This results in the retention of more of a person's uncertainty and results in elicited ranges capturing the true value more often—significantly reducing over-precision (chapter 7) and does so without any need for the elicitee to understand why it works. That is, it does not require the person using it to know what overconfidence or anchoring are in order for them to benefit from its use.

12.3 Conclusions

While not coming close to being a comprehensive account of methods for reducing or avoiding bias, the above discussion will, I hope, serve as a primer on the different approaches that can be taken by those interested in reducing biases in the decisions that they and those around them make. As noted above, while awareness is not sufficient to eliminate bias, it is generally necessary (except where this is obviated by the presence of an expert elicitor or elicitation tool). That is, knowledge of the biases allows you to consider ways that they might affect the decisions that you and others make and, thus, to take countermeasures or make allowances for their effect on judgements—improving your decisions and (on average) your outcomes. This can be achieved in a number of ways. Such knowledge can allow you to: avoid a bias entirely; recognise the conscious actions that might reduce it; see the value in a change in perspective from the inside (specific) to the outside (distributional) view of events; enable you to redesign decision and elicitation processes so as to limit the effect of particular biases; or prompt you to utilise more formal decision analysis techniques or expert elicitors.

All of these have the potential to reduce the impact of biases on the decisions that we make although, given the range of biases to which we are subject, it may be unrealistic to expect to be able to eliminate them entirely. Rather, understanding the nature of biases and how they come about is key to identifying *which* biases are likely to cause the greatest problems in different types of decisions and, thus, which debiasing strategies to consider. For example, if your concern is regarding potential gender or racial bias in hiring decisions, then you might consider combining a 'semi'-blind review of applicant CVs (with the candidate's name removed throughout) with an MCDM approach to ensure that they are all are judged on the same criteria. In contrast, for a situation where parameter estimates are being elicited for use in modelling, retaining an appropriate amount of uncertainty might be considered key and, thus, an approach like SPIES or the MOLE might be sufficient. Finally, for important, complex decisions, it might be worth engaging an expert elicitor to ensure that as many biases as possible are avoided.

By way of example, in the following chapter, we will consider a specific problem faced by scientists and science communicators, that of false beliefs, which results from the biases discussed in previous chapters. We consider how our understanding of human cognitive processes can help us to understand why people form these beliefs, why they can be so hard to displace or remove and how we can best approach this task.

References

Apgar V 1952 A proposal for a new method of evaluation of the newborn *Class. Pap. Crit. Care* **32** 97

Bratvold R and Begg S 2010 *Making Good Decisions* (Richmond, TX: Society of Petroleum Engineers)

Grove W M, Zald D H, Lebow B S, Snitz B E and Nelson C 2000 Clinical versus mechanical prediction: a meta-analysis *Psychol. Assess.* **12** 19

Haran U, Moore D A and Morewedge C K 2010 A simple remedy for overprecision in judgment *Judgm. Decis. Making* **5** 467

Hawkins J T, Coopersmith E M and Cunningham P C 2002 Improving stochastic evaluations using objective data analysis and expert interviewing techniques *Society of Petroleum Engineers Annual Technical Conference and Exhibition*

Kahneman D and Tversky A 1982 Variants of uncertainty *Cognition* **11** 143–57

Kuncel N R, Klieger D M, Connelly B S and Ones D S 2013 Mechanical versus clinical data combination in selection and admissions decisions: a meta-analysis *J. Appl. Psychol.* **98** 1060–72

Lichtenstein S, Fischhoff B and Phillips L 1982 Calibration of probabilities: the state of the art to 1980 *Judgement Under Uncertainty: Heuristics and Biases* ed D Kahneman, P Slovic and A Tverski (Cambridge: Cambridge University Press)

Meehl P E 1954 *Clinical Versus Statistical Prediction: A Theoretical Analysis and Review of the Evidence* (Minneapolis, MN: University of Minnesota Press)

Morgan M G and Henrion M 1990 *Uncertainty: A Guide to Dealing with Uncertainty in Quantitative Risk and Policy Analysis* (New York: Cambridge University Press)

Mussweiler T, Strack F and Pfeiffer T 2000 Overcoming the inevitable anchoring effect: considering the opposite compensates for selective accessibility *Person. Social Psychol. Bull.* **26** 1142–50

Russo J E and Schoemaker P J 1992 Managing overconfidence *Sloan Manag. Rev.* **33** 7

Spetzler C S and Stael von Holstein C A S 1975 Probability encoding in decision analysis *Manag. Sci.* **22** 340–58

Welsh M and Begg S 2018 More-or-less elicitation (MOLE): reducing bias in range estimation and forecasting EURO *J. Decis. Process.* accepted

Welsh M B, Begg S H and Bratvold R B 2006 SPE 102188: correcting common errors in probabilistic evaluations: efficacy of debiasing *Society of Petroleum Engineers 82nd Annual Technical Conference and Exhibition (Dallas, TX)*

Welsh M, Lee M and Begg S 2009 Repeated judgments in elicitation tasks: efficacy of the MOLE method *Proc. 31st Ann. Conf. Cognitive Science Society* ed N A Taatgen and H van Rijn (Austin, TX: Cognitive Science Society) pp 1529–34

Welsh M, Lee M and Begg S 2008 More-or-less elicitation (MOLE): testing a heuristic elicitation method *Proc. 30th Ann. Conf. Cognitive Science Society* ed V Sloutsky, B Love and K McRae (Austin, TX: Cognitive Science Society) pp 493–8

Wilson T D, Houston C E, Etling K M and Brekke N 1996 A new look at anchoring effects: basic anchoring and its antecedents *J. Exp. Psychol.* **125** 387

Yaniv I and Foster D P 1995 Graininess of judgment under uncertainty: an accuracy-informativeness trade-off *J. Exp. Psychol.* **124** 424

IOP Publishing

Bias in Science and Communication
A field guide
Matthew Welsh

Chapter 13

A river in Egypt: denial, scepticism and debunking false beliefs

This chapter examines how and why people end up with biased viewpoints—particularly in cases when contrary scientific evidence already exists. This includes a discussion of best practices for communicating information in order to avoid bias, as well as a discussion of how to salvage situations where bias has already taken hold in the form of firmly held misconceptions. That is, how denials, myths or false beliefs come about and how best to debunk them.

13.1 Facts, factoids and fictions

False beliefs are common across all areas of human enterprise—so common, in fact, that they have their own name: factoids. Factoids are commonly repeated items of information that feel like facts yet are demonstrably untrue. That is, they are unsupported by—or even contrary to evidence—and yet continue to circulate amongst society and be believed by large numbers of people. This is particularly true for scientific factoids—which borrow the credibility of science to bolster their own believability—but, perhaps more concerningly, factoids can be so well-accepted that even some scientists accept them as facts. For example, a study by Herculano-Houzel (2002) found that 72% of lay-people believed the factoid that people only use 10% of their brain (a myth that has no basis in neuroscience or psychology and no clear provenance; Boyd 2008) but that 6% of neuroscientists did as well—with a further 28% not being sure whether it was true or not.

Given that the role of scientists is to uncover the truth about how the world works and then share this information with others, understanding how and why these factoids spread—seemingly so much faster that actual facts—is an important undertaking. Both because, in an ideal world, we would stop falsehoods being circulated but also want to better understand how to increase the propagation rate of genuine, scientifically supported facts.

13.1.1 If it does no harm?

Before this, however, we should consider whether the propagation of factoids and other false information is actually harmful. Does it, for example, matter if people think they should drink eight glasses of water per day in addition to any other fluids (coffee, tea, juices)? The answer in this case is: probably not. Having a few extra glasses of water is unlikely to cause more harm than an extra toilet run or two each day. What about someone who believes that the Earth is flat? Again, in most situations, this false belief has very little impact on their ability to function on a day-to-day basis. Other false beliefs, however, are less benign, resulting in changes to people's behaviours that can have significant consequences for their own health and the well-being of those around them.

In terms of recent, science-based information, perhaps the clearest examples of this are the 'debates' over the safety of vaccinations and over climate change. In both cases, the overwhelming scientific consensus is clear: vaccinations are a safe and effective means of reducing the risk to both individuals and society from contagious diseases; and the global temperature is rising as a result the extra carbon dioxide (and other greenhouses gases) that are produced as by-products of human fossil fuel use and other activities. Amongst the public—which for these purposes can be considered anyone outside of the specialities of immunology and climate science, however, these 'facts' are commonly disputed and factoids are used as ammunition in these battles.

For example, some 'climate sceptics' (wrongly) argue that, even if the climate is changing, humans could not be responsible—because (they claim) volcanos produce far more CO_2 than human activities. This, however, is a factoid. Comparisons of CO_2 outputs from volcanoes and human activities indicate that, in a typical year, the contribution of volcanoes to the increase in atmospheric CO_2 is less than 1% that of human activities (see, e.g. Gerlach 2011). In fact, Gerlach suggests that, even if we were to experience a 1-in-100 000-year super-eruption like Mount Toba (74 000 years ago) *every* year, volcanoes would still add less CO_2 to the atmosphere than human activity.

In the realm of vaccinations, evidence is similarly one-sided. Vaccines for diseases like measles very rarely cause dangerous side-effects (less than a 1-in-1 000 000 chance; CDC 2012) and greatly reduce the chance of a person contracting the disease—which, itself, results in hospitalisation in more than 25% of cases and a 1-in-500 chance of death (alongside potential complications including deafness and brain damage; CDC 2015). A significant number of parents, however, believe that the measles–mumps–rubella (MMR) vaccine has significant side-effects including the potential to cause autism—as a result of a completely discredited but widely reported paper by Wakefield *et al* (1998 [retracted]). Despite immediate reactions from other scientists pointing out the flaws in the paper, followed by repeated failures to replicate the work and, finally, evidence of deliberately fraudulent and misleading practices by the authors (see, e.g. Godlee *et al* 2011)—which led to the paper being retracted by the publishing journal and the lead author struck off the medical register in the UK—its effect on vaccination rates was marked.

From a high of 92% prior to the Wakefield paper's publication, vaccination rates in the UK fell to around 80% in 2003–2004 (Gupta *et al* 2005)—and remain below the recommended 90% 'herd immunity' level—which has resulted in a mumps epidemic in 2005 and measles being declared endemic in the UK population in 2008 for the first time since 1994 (EuroSurveillance Editorial Team 2008).

So, the spread of factoids and falsehoods can be a serious problem with significant costs to society and individuals. That being the case, we need to understand how and why this occurs and what we can do about it.

13.1.2 Gullibility is good

So, why do people so easily believe false information? At a fundamental level, the answer is: because people are gullible. This may sound like a harsh indictment of humanity but is, in fact, simply a reflection of the type of creatures we are: social creatures born with only limited capabilities and who need to learn the majority of information and behaviours required to function in the world. As a result, our primary source of information about the world is often not first-hand experience but rather what we are told: by parents, teachers and peers.

Given this, high levels of scepticism would actually be counter-productive. It makes more sense to assume that people are giving you relevant, factually correct information and only question that to the extent that what you are being told fails to mesh with other things you believe. When told something new that fits into whatever worldview you have built up so far, accepting it is likely to be the better option in most cases. In fact, studies of human conversational norms show exactly this—with people expecting statements to be truthful, relevant and informative (see, e.g. Grice 1975). It also helps to explain the differences in culture and tradition seen across the peoples of the world. For example, while religiosity or belief in the supernatural is common across cultures (perhaps as a result of inherent aspects of human cognitive processes; see, e.g. Azar 2010), the particular religion or set of beliefs that an individual holds is largely dependent on the religion of their parents and the larger society in which they have been raised—that is, what they have been told about the supernatural.

This implies that 'gullibility' is a survival trait. This is true both in terms of acquiring shared cultural knowledge, which can improve social cohesion, and given the requirements of social intelligence—where specialist knowledge is shared amongst a variety of society members and no single person can be expected to have the time to learn everything they need to function in society via trial and error.

As a result, as will be outlined in more detail below, unless there are clear indicators that the speaker is trying to mislead, a new piece of information will tend to be accepted unless its acceptance threatens or contradicts already existing knowledge.

13.2 Boosting believability

There are, of course, a number of things that alter how inherently believable a statement is. Three of these that relate back to various aspects of decision making and bias discussed earlier in the book are expanded on below. Specifically: the coherence of the story; how often and from how many people you hear it; and the trustworthiness of the source.

13.2.1 That makes sense

As noted above, people's default position is to accept new information—so long as it fits into their pre-existing beliefs or, at least, does not immediately prompt recognition that such a mismatch exists. (This caveat is added because it is perfectly possible for people to hold multiple, logically inconsistent beliefs—so long as they do not realise that they are inconsistent. It may be that only when a belief is examined logically that everything it entails becomes clear and the person begins to experience 'cognitive dissonance'—the feeling of discomfort from knowingly holding contradictory beliefs (Festinger 1957) that prompts people to revise their beliefs or otherwise act to reduce the dissonance.)

Of course, this means that one way of avoiding this difficulty is to accept new information uncritically—that is without consideration of its consequences for your other beliefs. In effect, believing it without assessing how well it accords with other information. Bensley and Lilienfield's (2017) paper on misconceptions about psychology offer supports for this idea. Specifically, they found that people who held the most misconceptions were those who had the highest scores on a faith in intuition scale and who scored lowest on a measure of critical thinking. (Interestingly, the results also showed the Dunning–Kruger effect (see chapter 7), with the people holding the most misconceptions also being the most overconfident in terms of how many questions they predicted they would get right.)

This implies that misconceptions or factoids, being wrong, are more likely to trigger dissonance when critically examined—as this will increase the likelihood of the person finding a mismatch with their other beliefs and thus rejecting the factoid as inconsistent. This distinction maps onto the two systems approach to decision making discussed in chapter 2. As noted throughout this book, the heuristics and biases approach aligns with this two systems approach and argues that in system 1, intuitive processes are implicated in susceptibility to a range of biases—and this work suggests that acceptance of false beliefs may be amongst these.

Another key attribute relating to how believable a new fact (or factoid) is, is the clarity or coherence of the causal story into which it fits, as raised in chapter 8. For example, if the new 'fact' explains something that the person did not, previously, have an explanation for, then it is more likely to be believed. This, it seems, explains a significant part of why the supposed link between autism and the MMR vaccine proves so seductive to parents. In this case we have parents who are told by doctors and scientists that there is no simple explanation for what causes autism. This, combined with the fact that autism is often first diagnosed at around the age when the vaccine is administered leaves an explanatory gap—waiting to be filled by any proposed cause. In the absence of any clear, alternative explanation, this factoid can be inserted into the causal structure without contradicting any current information, meaning that it is even likely to escape being flagged if a person's critical thinking is activated as described above.

Finally, it is worth remembering the maxim easy = true, from chapter 9. While the link between the MMR vaccine and autism is not true, the causal structure is very easily understood. By comparison, the complex science around other potential causes of autism is very hard to understand. Given that people are more likely to accept simple explanations than complex ones, when communicating to lay-people, simple explanations are not just more likely to be understood but also more likely to be believed. This sheds light on how press releases relating to scientific findings should and should not be drafted—in order to ensure that the core message is believed while avoiding the creation of false beliefs through oversimplification.

13.2.2 Echo...

The second major aspect of believability is repetition—the more often a fact or factoid is repeated, the more likely it is to be believed. This, of course, makes sense in the social learning context described above—if everyone around you is saying the same thing, then it makes even more sense to accept that statement as true rather than questioning it (Festinger 1954). This can cause problems, however, as amount of repetition is not always informative regarding the source of the information. For example, if six different microbiologists tell you that there is evidence that a new antibiotic is effective against Methicillin-resistant *Staphylococcus aureus* (MRSA), it matters whether those six colleagues have independently corroborated this or simply all read the same, single study. The amount of repetition, however, does not tell you this and repetition alone is sufficient to make people more inclined to believe statements—because familiarity makes a statement easier to process as per the fluency effect discussed in chapter 9. In fact, there is evidence that people do not always retain the association between the source of information and the information itself, with the result that even hearing the same belief from the *same* source repeatedly can result in them recalling that belief as being widely held (Weaver *et al* 2007).

This doubtless has had an effect on the spread of scientific myths like the link between MMR vaccine and autism—as the claim was seized on and repeated by a large number of media outlets—over several years, in fact, as Wakefield was given airtime to promote his claims and argue against those disagreed with him—and then propagated onto countless websites. Thus, everyone has probably heard the claim that vaccines cause autism from 'multiple' sources, increasing its intrinsic believability.

The effect of repetition on believability is also an argument against 'balanced' reporting in journalism, where balanced is taken to mean providing equal time to both sides of a debate—no matter how few people there are on one side or how little evidence supports it. This can result in fringe views being repeated just as often as mainstream views with the result that people consuming the media have no indication which *is* the mainstream view and, as a result, being left with the impression that the scientific community is divided as, when they think back to how many pieces of evidence they have for each side (i.e. use the availability heuristic; chapter 8) they find equal numbers. This was certainly the case with the vaccine–autism 'debate' (Clarke 2008) and also in the case of climate science, where the traditional journalistic approach of inviting one climate scientist and one climate change 'sceptic' gives the impression of an active, scientific debate when, in fact, research (Doran and Zimmerman 2009) shows the majority of earth scientists agree on the basic facts—that the world is warming (90%) and that human activity is a significant contributor to this (86%)—and that these numbers are higher for climate scientists who actively conduct research into whether the climate is changing (>95% for both questions, with ~2% disagreeing and ~2% being unsure). This implies that to give a truly balanced viewpoint from the point of view of available instances, for every minute of time given to a 'sceptic', 10–20 min would need to be given to the consensus view.

This is, of course, practically difficult for journalists and this makes it easy for people to take advantage of this to drive their own agendas—with unscrupulous politicians or corporations able to push an alternative viewpoint in the knowledge that this will be discussed at greater length than it might deserve and thus muddy the water regarding the scientific consensus (as was done by cigarette companies for many years; see, e.g. Smith *et al* 2011).

It is also a particularly worrying tendency given the fragmentation of the media landscape, with the explosion of online sources enabling people to search for sources of information that accord with their pre-existing view-points (i.e. confirmation bias as described in chapter 5) and finding a plethora of voices that repeating those views back to them—leading to the false consensus bias (chapter 5), where people believe that their views are in the majority simply because they have entered an environment where other voices cannot be heard. This has been called an 'echo chamber' effect, where even fringe views can be repeated often enough within a closed environment that people come to not only believe them but believe that they are mainstream.

13.2.3 ...and Narcissus

The third major area of bias that leads to false beliefs being accepted and propagated relates to trust. That is, who do you trust? Obviously, some sources of information are more trusted than others and should, at least, be more trusted in different domains. For example, as a child, your trust in your parents' opinions was likely very strong but, as an adult, you have probably realised that their knowledge has gaps like anyone else and so, as a result, when you want to ask someone about quantum effects, for example, there are probably people you would trust more (with apologies to any quantum physicist parents).

The reasons that we trust or distrust people, however, can be far from unbiased. While one would expect a reliability effect—whereby we are more inclined to trust people who say things that are true, we need to remember that, as discussed above, new pieces of information may only trigger a check for their truth if they contradict something that we already believe. As such, in areas where a lay person might, quite reasonably, have no opinion (including climate change or vaccine safety) the truthfulness of the information is unlikely to be questioned. Instead, people are likely to make their decisions regarding whether they should trust this new information based on the personal characteristics of the messenger—which brings with it all of the problems of stereotyping and halo effects discussed in chapter 9. That is, we are more likely to trust information given to us by people who seem more like us because, like Narcissus, we tend to hold positive views of ourselves in terms of attractiveness, ability and morality (an example of overplacement, as discussed in chapter 7). Then, to the extent that another person seems like us, we assume that they will, likewise, share these positive traits (the halo effect, chapter 9).

So, I know that I am trustworthy and you remind me of me. Therefore, I assume that you are trustworthy too—and uncritically accept the information you are offering.

This, however, works in reverse too: if you seem different to me, then I am less likely to uncritically accept what you say. So, if I am an Australian, rural-dwelling lay-person and you are a European, urban-dwelling scientist, for example, I may see very little correspondence between our characteristics and very little overlap in our worldviews. As a result, I may assume that you do not share my traits of honesty and integrity—making me less likely to believe you.

The implication of this is probably uncomfortable for scientists but, nevertheless, needs to be considered when attempting to convey scientific information: the messenger matters. While we might hope that the message itself is sufficient to convince people of its truth, this is probably not the case and, when communicating with a particular audience, understanding the characteristics of that audience and tailoring the message to them will increase the likelihood of its uptake. For example, Kahan (2010) argues people are particularly resistance to information that challenges their worldview and thus that to increase acceptance of information, one needs to affirm the listener's worldview, couch the information in terms that accord with that worldview (e.g. talking about climate change as an opportunity for some industries rather than as a threat), and have it delivered by someone who seems to hold the same worldview—or at least a diverse set of people. In this way, accurate information is more likely to be accepted by listeners. Pitting scientific findings against a person's world-view, in contrast, is likely to result in them dismissing science or its ability to meaningfully tackle questions in areas related to the person's worldview (the 'scientific impotence excuse'; Munro 2010).

13.3 Fool me once, fool me forever?

One, initially confusing, aspect of false beliefs is their degree of resistance to change. Once a person has taken on a new piece on information, simply telling them that it is wrong tends not to change their mind back its pre-falsehood state—as was seen in the data relating to vaccination rates in the sections above. Even after the Wakefield study was debunked and withdrawn, vaccination rates remain below prior levels and some fraction of the population retained the false belief that vaccines cause autism.

The reason for this, however, is fairly straightforward. As noted above, a layperson is, initially, unlikely to have a causal explanation for autism at all. Thus, there is an open position in their causal schema into which the false belief neatly slots, completing the schema and making a coherent story. When a retraction is attempted, however, the retraction attempts to replace the false belief with its own negation. This means, however, that the retraction most definitely results in a mismatch with the person's current worldview—as this includes the false belief. As noted above, this will tend to trigger a person's scepticism and, as a result, they are likely to assess the *retraction* more critically and be less likely to accept it.

Additionally, this believability contest between the false belief and the retraction leaves the person with two alternatives—to return to the unsatisfying and incoherent structure where the cause of autism is not known or to retain the false belief in spite of the retraction—as shown in figure 13.1. Given this, it is perhaps not very surprising that retractions alone are not an effective means of reducing false beliefs and sometimes result in 'backfires' where the person's commitment to the false belief is actually increased (for a review, see Lewandowsky *et al* 2012).

Even if one draws on the fact that repetition aids in the acceptance of true facts as well as false ones and thus attempts repeated retractions or refutations of a false

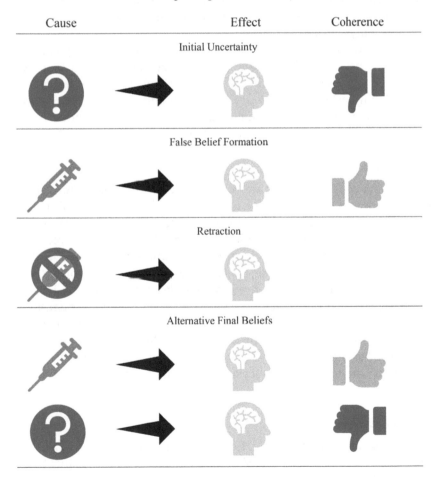

Figure 13.1. Advantage of first cause in false belief formation. Note that the false belief that the vaccine causes the developmental issues replaces the intrinsically incoherent initial state of uncertainty with a coherent (if wrong) explanation. The retraction not only has to try to replace the existing cause but, in doing so, reduces the coherence of the explanation. As a result, the false belief may be retained.

belief, while beneficial, this tends not to be entirely successful (see, e.g. Ecker *et al* 2011)—seemingly because of this coherence gap.

13.4 Debunking

The above describes how various of the biases encountered in previous chapter combine in the real world to aid the spread and acceptance of false beliefs that have proven to be robust to simple corrections. While one might hope that improved education would reduce this problem—with more educated people being less affected by false beliefs—in fact, the impact of worldview seems to interact with level of education in unhelpful ways. As an example, Hamilton (2011) showed that, amongst Democrats, level of education correlates *positively* with perception of threat from climate change but, amongst Republicans the correlation is *negative*. This suggests that more educated people, rather than being immune to false beliefs may, instead, be more adept at seeking out information that supports their worldview (confirmation bias, chapter 5).

So, given the robust strength of false beliefs and the costs they can impose on individuals and societies, it is important to consider the ways in which they can be undermined or replaced and some of the problems that can result—using the terminology from Lewandowsky *et al* (2012) as an outline. (For those interested in a longer exposition of the problem and possible solutions, see Lewandowsky *et al* 2012 and Cook and Lewandowsky 2011, respectively.)

13.4.1 Continued influence

This refers to the continued influence of a false belief after an attempted retraction. That is, where people still rely on the false belief. This seems to result from the lack of an alternative explanation, which still leaves the false belief as the most coherent account (as per figure 13.1). Lewandowsky *et al* (2012) therefore recommend the provision of an alternative account as the best approach—paired with repeated retractions to weaken the effect of the false belief. That is, if you intend to stop someone believing something, you need to not only explain why that belief is wrong but also provide an alternative fact to replace this belief in whatever causal schema it serves. For example, to convince someone that MMR vaccine does not cause autism, one needs to not only explain the facts regarding the lack of any relationship but also, to the best of your ability, provide an explanation—if not for the cause of autism then for the temporal co-occurrence of the vaccine and first diagnosis of autism.

13.4.2 Familiarity backfire

This effect describes the tendency for attempted refutations of false beliefs to sometimes increase a person's belief in it. This occurs because, in attempting to refute the myth, people often start by repeating the myth. Not only does this additional repetition increase fluency and thus the believability of the myth (fluency, chapter 9) but the fact that the myth is presented first means that it is the part of the explanation that is most likely to be recalled (primacy effect, chapter 8). Thus, the myth-then-refutation structure can result in people only recalling another repetition of the myth. To combat this, Lewandowsky et al (2012) suggest re-ordering retractions and refutations to place the correct facts first and then segue from them to a discussion of the false belief—after warning people that you will be presenting them with this false information. This ensures that a person's memory has the greatest chance of recalling the true information and, at the same time, primes them to think more critically about the false belief when it is presented—which increases the likelihood of the person recognising a mismatch between the false belief and other aspects of their knowledge.

13.4.3 Overkill backfire

An overkill backfire occurs when, in attempting to refute a false belief, too much information is presented, with the result that the refutation becomes far more complex than the false belief that it is trying to displace. As noted in chapter 9, however, people's preference is for simple explanations. Thus, in a competition between a simple, false belief and a complex refutation, people tend to hold to the false belief. Lewandowsky et al's (2012) advice for this is, of course, to make rebuttals as brief as possible—concentrating on a few arguments against the false belief's truth rather than trying to include all possible evidence of its falsehood. In addition to this, they recommend attempting to foster 'healthy' scepticism. That is, drawing attention to, for example, any questionable provenance or source of the false belief so as to prompt people to examine it more critically.

13.4.4 Worldview backfire

The final debunking problem that Lewandowsky et al (2012) address is the worldview backfire, where an attempted refutation is judged to be threatening to or at odds with the listener's strongly held worldview. As a result, rather than accepting information that might dismantle the way in which they understand the world, they dismiss any counterevidence and seek out evidence that supports their initial (false) position. If evidence cannot be presented in a manner that avoids this problem, a way around this is to explicitly affirm the listener's worldview—focussing on opportunities rather than threats, for example, or even by having the listener self-affirm their personal values—both of which tend to increase their receptivity to evidence.

13.4.5 Choice architecture

An alternative raised by Lewandowsky et al (2012), which relates back to the paternalistic liberalism ('nudges') approach discussed in chapter 6, is to avoid confrontation with people's worldviews entirely by focusing on the development of choice architecture that prompts people to make decisions in line with scientific evidence—rather than trying to convince them of the scientific evidence itself. This does, however, require significant understanding of your audience in order to understand how choices should be structured in order to promote the optimal decision. For instance, while changes in the description of a natural disaster from percentage terms to natural frequency terms prompts stronger reactions from listeners (Welsh *et al* 2016), the nature of this reaction depends on the listeners' prior experience with different types of natural disaster. Specifically, with the change in format, those with little experience of bushfires tended to be prompted to sell their house, whereas those with more experience tended to opt for more disaster-preparation. Given this, choice architecture is an area that requires much more research.

13.5 Conclusions

The above makes abundantly clear the need for scientists to, more centrally, consider the role of communication in science—for the betterment of society. While, in an ideal world, 'the truth would out' and false beliefs would be swiftly corrected, the research and examples presented above demonstrate the difficulties faced by scientists seeking to prevent the propagation of false beliefs and uproot them where they exist. This is because the cognitive tendencies and limitations discussed in previous chapters act together to aid in the acceptance of false beliefs and, then, to resist their replacement.

In addition to its many benefits, the democratisation of information that has resulted from the growth and spread of the internet has, unfortunately, brought with it an increased ability of falsehoods to spread and for people to engage in uncritical 'cherry picking' of information in order to support almost any position. This results in a greater need for clarity in scientific communications—both of novel findings and the presentation of consensus views designed to combat misinformation.

While this task is daunting, with an understanding of how and why people succumb to false beliefs and other biases in thinking we can equip ourselves with tools with which to weed the garden of knowledge.

References

Azar B 2010 A reason to believe *Monit. Psychol.* **41** 52–6

Bensley D A and Lilienfeld S O 2017 Psychological misconceptions: recent scientific advances and unresolved issues *Curr. Dir. Psychol. Sci.* **26** 377–82

Boyd R 2008 Do people only use 10 percent of their brains? *Sci. Am.* 7 February www.scientificamerican.com/article/do-people-only-use-10-percent-of-their-brains/ (Accessed: 30 December 2017)

CDC 2012 *MMR Vaccine Information* Sheet Center for Disease Control www.cdc.gov/vaccines/hcp/vis/vis-statements/mmr.html (Accessed: 30 December 2017)

CDC 2015 *The Pink Book. Chapter* 13, *Measles* Center for Disease Control. www.cdc.gov/vaccines/pubs/pinkbook/meas.html (Accessed: 30 December 2017)

Clarke C E 2008 A question of balance: the autism–vaccine controversy in the British and American elite press *Sci. Commun.* **30** 77–107

Cook J and Lewandowsky S 2011 *The Debunking Handbook* www.skepticalscience.com/docs/Debunking_Handbook.pdf (Accessed: 31 December 2017)

Doran P T and Zimmerman M K 2009 Examining the scientific consensus on climate change *Eos Trans. Am. Geophys. Union* **90** 22–3

Ecker U K, Lewandowsky S, Swire B and Chang D 2011 Correcting false information in memory: manipulating the strength of misinformation encoding and its retraction *Psychonom. Bull. Rev.* **18** 570–8

EuroSurveillance Editorial Team 2008 Measles once again endemic in the United Kingdom *EuroSurveillance* **13** 18919

Festinger L 1954 A theory of social comparison processes *Hum. Relat.* **7** 117–40

Festinger L 1957 *A Theory of Cognitive Dissonance* (California: Stanford University Press)

Gerlach T 2011 Volcanic versus anthropogenic carbon dioxide *Eos Trans. Am. Geophys. Union* **92** 201–2

Godlee F, Smith J and Marcovitch H 2011 Wakefield's article linking MMR vaccine and autism was fraudulent *Bri. Med. J.* **342** c7452

Grice H P 1975 Logic and conversation *Syntax and Semantics* vol 3 ed P Cole and J Morgan (New York: Academic) pp 41–58

Gupta R K, Best J and MacMahon E 2005 Mumps and the UK epidemic 2005 *Bri. Med. J.* **330** 1132–5

Hamilton L C 2011 Education, politics and opinions about climate change evidence for interaction effects *Climatic Change* **104** 231–42

Herculano-Houzel S 2002 Do you know your brain? A survey on public neuroscience literacy at the closing of the decade of the brain *Neuroscientist* **8** 98–110

Kahan D 2010 Fixing the communications failure *Nature* **463** 296–7

Lewandowsky S, Ecker U K, Seifert C M, Schwarz N and Cook J 2012 Misinformation and its correction: continued influence and successful debiasing *Psychol. Sci. Publ. Interest* **13** 106–31

Munro G D 2010 The scientific impotence excuse: discounting belief-threatening scientific abstracts *J. Appl. Soc. Psychol.* **40** 579–600

Smith P, Bansal-Travers M, O'Connor R, Brown A, Banthin C, Guardino-Colket S and Cummings K M 2011 Correcting over 50 years of tobacco industry misinformation *Am. J. Prevent. Med.* **40** 690–8

Wakefield A J *et al* 1998 Ileal lymphoid nodular hyperplasia, non-specific colitis, and pervasive developmental disorder in children *Lancet* **351** 637–41

Weaver K, Garcia S M, Schwarz N and Miller D T 2007 Inferring the popularity of an opinion from its familiarity: a repetitive voice can sound like a chorus *J. Pers. Soc. Psychol.* **92** 821

Welsh M, Steacy S, Begg S and Navarro D 2016 A tale of two disasters: biases in risk communication *Proc. 38th Annual Meeting of the Cognitive Science Society* ed A Papafragou *et al* (Philadelphia, PA: Cognitive Science Society) pp 544–9

Part IV

Conclusions

Bias in Science and Communication
A field guide
Matthew Welsh

Chapter 14

The field guide: general conclusion and spotters guide to biases

The previous chapters have attempted to explain when, how and why biases affect people's decision making, with a focus on decisions relevant to the pursuit and communication of science. The message I hope you will take away is that 'biased' is the natural state of human decision making, simply because we are attempting to use limited cognitive processing power and memory capacity to interpret information from an extremely complex world. Moreover, a world that, as a result of advances in communications technology, no longer shares the same information characteristics as the natural environments in which our brains and cognition evolved. These mismatches can, therefore, amplify the tendency of some heuristics to cause bias.

As discussed in chapters 11–13, biases can also interact so as produce problems like the spread of false beliefs that have significant societal impact but are not obviously the result of a single bias. As was the case for complex, important decisions, these need to be tackled with more complex approaches—such as the expert interview debiasing strategies described in chapters 12 and the debunking strategy described in chapter 13 that take into account these multiple potential causes of bias.

The good news for people interested in reducing the role that biases play in science and communication, however, is that people's limitations are shared. That is, to a first-order approximation, everyone uses the same cognitive processes and, therefore, tends to be affected by the same biases. While this may not, immediately, seem like a good thing, it is to the extent that this makes biases systematic and predictable in terms of when and how they affect decisions. Thus, as discussed in many of the previous chapters, we can recognise those situations and set in place processes to limit or make allowances for the effects of these biases.

Given the range of biases discussed and the overlapping situations in which they can occur, it is too much to hope that the knowledge gained from this book (or even a far more detailed and comprehensive reading of the still-developing heuristics and biases literature) will allow a reader to recognise all of the biases that might affect a particular decision but it will, I hope, enable the recognition of those most likely to have a large effect in a particular circumstance and suggest some ways of reducing their effects.

14.1 Spotter's guide to bias

To this end, this, final chapter is structured as a 'field guide'—with table 14.1 detailing a set of the most recognisable biases, the location of their fuller descriptions in this book, examples of these and the environments in which they occur, and steps that can help to reduce their impact.

Table 14.1. Spotter's guide to biases.

Bias	Occurrence	Mitigation
Anchoring (chapter 8): Starting numerical estimation process from any available number.		
Estimates tend to be too close to the anchoring value.	Any time someone makes a numerical estimate under uncertainty.	Deliberate introduction of multiple possible starting points for an estimate.
Availability (chapter 8): Basing judgements of likelihood on recalled number of events.		
Inaccuracy of memory due to personal salience, media coverage bias, etc, result in misjudgement of likelihoods.	Contributes to the planning fallacy and overestimation of likelihood of rare but memorable events.	Unpack general categories into specific ones prior to estimation.
Base rate neglect (chapter 4): Focus on diagnostic evidence, ignoring prior occurrence rates.		
Reliance on new data ignores the impact of false positives and can overstate the actual updated probability.	When updating probabilities given new evidence.	Use of Bayes theorem or probability trees.
Confirmation bias (chapter 5): Searching for evidence that confirms current beliefs.		
Results in disconfirming evidence being missed and too much confidence in own position.	Any time evidence is being searched for or evaluated to test a hypothesis—e.g. in peer review.	Adherence to the scientific method and reliance on falsification. Greater weighting of introduction and method in peer review.
Fluency (chapter 9): Ease of understanding increases believability.		
Commonly repeated statements seem truer and simpler explanations are preferred.	Repeated media presentation of a scientific claim can make it more believable. Simple, false explanations preferred to complex, accurate ones, leads to acceptance of scientific myths.	Increasing the difficulty of consuming the information prompts greater critical thought (e.g. blurry fonts, poor audio). Recall that simplicity must also be weighed against accuracy.

Framing (chapter 6): Options rephrased to sound like losses or gains change people's choices.

Positively framed options prompt risk aversion while negative ones prompt risk seeking.	In situations where riskier and more certain options are being compared (e.g. stock market versus bank investment).	Rewriting the same options in the opposite frame and seeking additional opinions to avoid hindsight bias.

Groupthink (chapter 9): Tendency of groups to quash dissent to produce agreement.

Suppression of non-conforming views produces overconfident, false consensus.	In homogenous, tight-knit groups with strong leaders.	Diverse groups and emphasis on critical evaluation of ideas. Generation of ideas outside of meetings.

Halo effect (chapter 9): Belief that positive (or negative) traits cluster together.

Results, e.g. in taller and more attractive people being presumed more competent.	When judging the characteristics of a person in the absence of good evidence —based on their looks or similarity to oneself.	Recall range truncation lessens the predictive power of these trait clusters. Seek actual evidence and avoid superficial information (e.g. blind review).

Outcome bias (chapter 11): Quality of decision being judged by outcome.

Good outcomes attributed to good decisions and bad outcomes to bad decisions despite evidence to contrary.	Publication bias—where negative results are attributed to poor methodology and positive ones to good methodology.	Judge decision quality in advance of outcomes (e.g. judge methods not results) and remember Russian roulette is a bad decision even if you win.

Overconfidence (chapter 7): Mismatch between accuracy and predicted accuracy.

Tendency of people's predictions and forecasts to be wrong more often than they expect.	When estimating point values or uncertain ranges for unknown parameters or future values. Judging how well you have performed or how well you compare to other people on some trait.	Awareness of overconfidence. Estimation practice with feedback. Use of elicitation tools (SPIES, MOLE) or reformatting of questions into formats that produce less bias.

Primacy and recency (chapter 6): Greater likelihood of recalling first and last objects of set.

Results in the order of presentation of options, evidence, etc, becoming important as the first one is most likely to be recalled— and then the last.	When lists of items need to be recalled. Or if evidence is presented for and against, the first point 'for' will have the greatest impact.	Use of Latin Square designs in research. Ensuring that presentation order matches the desired memorability— e.g. facts refuting a false belief should be presented before the belief.

(Continued)

14-3

Table 14.1. (*Continued*)

Bias	Occurrence	Mitigation
Sample size invariance (chapter 4): Overweighting of evidence from small samples.		
People assign the same weight to evidence (largely) irrespective of the quantity of data.	Relying on statistical significance when interpreting scientific papers without concern for sample size. Scientific results that are reported without mention of sample size.	Good statistics and recalling implications of the law of large numbers—that small samples are less reliable.
Stereotyping (chapter 9): Using group characteristics to predict individual behaviour.		
Results in inaccurate predictions of performance and behaviour. Confirmation bias causes behaviour to be interpreted in light of the stereotype.	When assessing characteristics of a person based on limited information, based on their group membership.	Seek individual information and avoid making decisions using superficial evidence, e.g. engage in blind review and use metrics rather than impressions to compare candidates.

14.2 Epilogue

Thank you for reading my book. I hope you have found it interesting—and that it proves useful to you in identifying biases in your own thinking as well as helping you to understand when and why other people might, even with the best of intentions, display bias in their thinking and communications.

Good luck with your bias spotting!

9 780750 313124